THE MESO...
175 MILLIO...

Millions of Years Before the Present	System / Period	Stage — EUROPE		Age — NORTH AMERICA
65	CRETACEOUS	MAASTRICHTIAN		GULFIAN
		CAMPANIAN		
		SANTONIAN		
		CONIACIAN		
		TURONIAN		
		CENOMANIAN		
100		ALBIAN		COMANCHEAN
		APTIAN	GARGASIAN / BEDOULIAN	
		BARREMIAN		
		NEOCOMIAN: HAUTERIVIAN		
		VALANGINIAN		
		BERRIASIAN		
140	JURASSIC	TITHONIAN		
		KIMMERIDGIAN		
		OXFORDIAN		
		CALLOVIAN		
		BATHONIAN		
		BAJOCIAN		
		AALENIAN		
		LIASSIC		
205	TRIASSIC	RHAETIAN		
		NORIAN		
		CARNIAN		
		LADINIAN		
		ANISIAN		
		SCYTHIAN		
240				

Rising from the Plains

———————

John McPhee

RISING

FROM THE

PLAINS

Farrar ᷓ Straus ᷓ Giroux

NEW YORK

Copyright © 1986 by John McPhee
All rights reserved
Printed in the United States of America
Published simultaneously in Canada by
Collins Publishers, Toronto
First edition, 1986
Library of Congress Cataloging-in-Publication Data
McPhee, John A.
Rising from the plains.
I. Title.

QE79.M29   1986   557.8   86-14891

Selections from the unpublished manuscripts and diaries
of E. W. Love copyright © 1986 by J. D. Love
and P. L. Holzinger
Map of Wyoming © 1986 by Tom Funk

The text of this book originally appeared in *The New Yorker*

Geological time scale adapted by Tom Funk from
F. W. B. van Eysinga's *Geological
Time Table* (Elsevier Scientific Publishing Company,
The Netherlands) and *A Geologic Time Scale*
(W. B. Harland et al., Cambridge University Press)
in conformity with publications of the United States
Geological Survey and the Geological Society of America.
These endpapers have been revised since the publication of
Basin and Range and *In Suspect Terrain* on the basis
of recent radiometric dating

For Yolanda Whitman

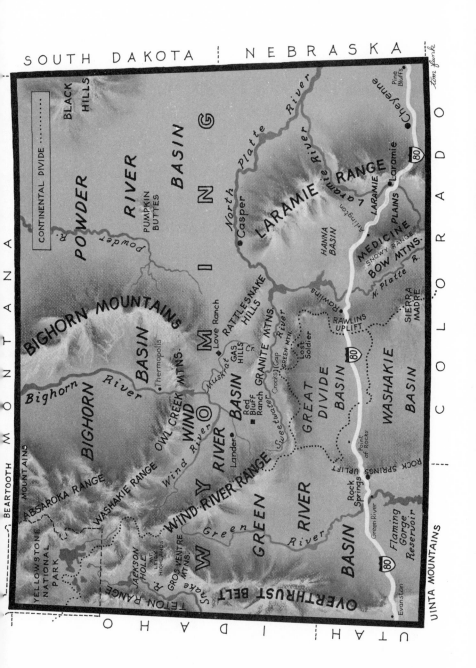

SOUTH DAKOTA | NEBRASKA

MONTANA

BEARTOOTH MOUNTAINS

COLORADO

UTAH | IDAHO

UINTA MOUNTAINS

Tom Funk

CONTINENTAL DIVIDE · · · · · · · · ·

BLACK HILLS

W Y O M I N G

POWDER RIVER BASIN

Powder R.

PUMPKIN BUTTES

North Platte River

Platte River

Laramie River

Cheyenne

Pine Bluffs

Laramie
Laramie

80

LARAMIE RANGE

Arlington

MEDICINE BOW MTNS.

Snowy Range

MEDICINE PLAINS

HANNA BASIN

N. Platte R.

Casper

BIGHORN MOUNTAINS

Love Ranch

RATTLESNAKE HILLS

GRANITE MTNS.

GAS HILLS

Cr.

Muskrat

Rawlins

RAWLINS UPLIFT

SIERRA MADRE

Bighorn River

Thermopolis

OWL CREEK MTNS.

WIND RIVER BASIN

Red Bluff Ranch

Crooks Gap

GREEN MTN.

Lost Soldier

GREAT DIVIDE BASIN

80

WASHAKIE BASIN

BIGHORN BASIN

WASHAKIE RANGE

Wind River

WIND O RIVER

Lander

Sweetwater

River

Point of Rocks

Rock Springs

ROCK SPRINGS UPLIFT

ABSAROKA RANGE

WIND RIVER RANGE

GREEN RIVER BASIN

Green River

River

Green River

Flaming Gorge Reservoir

80

Evanston

YELLOWSTONE NATIONAL PARK

JACKSON HOLE

Teton PASS

Leidy HIGHLANDS

GROS VENTRE MTNS.

TETON RANGE

Snake R.

OVERTHRUST BELT

C O L O R A D O

Rising from the Plains

———————

This is about high-country geology and a Rocky Mountain regional geologist. I raise that semaphore here at the start so no one will feel misled by an opening passage in which a slim young woman who is not in any sense a geologist steps down from a train in Rawlins, Wyoming, in order to go north by stagecoach into country that was still very much the Old West. She arrived in the autumn of 1905, when she was twenty-three. Her hair was so blond it looked white. In Massachusetts, a few months before, she had graduated from Wellesley College and had been awarded a Phi Beta Kappa key, which now hung from a chain around her neck. Her field was classical studies. In addition to her skills in Latin and Greek, she could handle a horse expertly, but never had she made a journey into a region so remote as the one that lay before her.

Meanwhile, Rawlins surprised her: Rawlins, where shootings had once been so frequent that there seemed to be—as citizens put it—"a man for breakfast every morning"; Rawlins, halfway across a state that was spending per annum far more to kill wolves and coyotes than to support its nineteen-year-old university. She had expected a "backward" town, a "frontier" town, a street full of badmen like Big Nose George, the road agent, the plunderer of stagecoaches, who signed his hidden-treasure maps "B. N. George." Instead, this October evening, she was met at the station by a lackey with a handcart, who wheeled her luggage to the Ferris Hotel. A bellboy took over, his chest a constellation of buttons. The place was three stories high, and cozy with steam heat. The lights were electric. There were lace curtains. What does it matter, she reflected, if the pitchers lack spouts?

☙ One spring day about three-quarters of a century later, a four-wheel-drive Bronco approached Rawlins from the east on Interstate 80. At the wheel was David Love, of the United States Geological Survey, supervisor of the Survey's environmental branch in Laramie, and —to an extent unusual at the highest levels of the science—an autochthonous geologist. The term refers to rock that has not moved. Love was born in the center of Wyoming in 1913, and grew up on an isolated ranch, where he was educated mainly by his mother. To be sure, experience had come to him beyond the borders—

a Yale Ph.D., explorations for oil in the southern Appalachians and the midcontinent—but his career had been accomplished almost wholly in his home terrain. For several decades now, he had been regarded by colleagues as one of the two or three most influential field geologists in the Survey, and, in recent time, inevitably, as "the grand old man of Rocky Mountain geology." The grand old man had a full thatch of white hair, and crow's feet around pale-blue eyes. He wore old gray boots with broken laces, brown canvas trousers, and a jacket made of horsehide. Between his hips was a brass belt buckle of the sort that suggests a conveyor. Ambiguously, it was scrolled with the word "LOVE." On his head was a two-gallon Stetson, with a braided-horsehair band. He wore trifocals. There was stratigraphy even in his glasses.

A remarkably broad geologist, he had worked on everything from geochemistry to structural geology, environmental geology to Pleistocene geology, stratigraphy to areal geology and mapping—and he had published extensively in all these fields. In the Bronco, he seemed confined—a restlessness that derived from a lifetime of travel on foot or horseback. He was taking me across Wyoming, at my request, looking at the rock in roadcuts of the interstate, which in seasons that followed would serve as portals for long digressions elsewhere in Wyoming in pursuit of the geologies the roadcuts represented. Once, in the Bighorn Basin, as we were rolling out our sleeping bags, I asked him what portion of the nights of his life he had spent out under the stars, and he answered, "One-third." A few minutes later, half asleep, he added a correction: "Let's say one-

[5]

quarter. I want to be careful not to exaggerate." He rolled over and was gone for the night. I passed out more slowly, while my brain tumbled heavily with calculation. Love was about seventy, and this, I figured, was something like his six-thousandth night on the ground. Well, not precisely on the ground. One must be careful not to exaggerate. He has had the same old U.S.G.S. air mattress for forty years. When it was quite new, it sprang a leak. He poured evaporated milk in through the valve and stopped the leak.

Now, as we crossed the North Platte River and ran on toward Rawlins in May, over the road were veils of blowing snow. This was Wyoming, not some nice mild place like Baffin Island—Wyoming, a landlocked Spitsbergen—and gently, almost imperceptibly, we were climbing. The snow did not obscure the structure. We were running above—and, in the roadcuts, among—strata that were leaning toward us, strata that were influenced by the Rawlins Uplift, which could be regarded as a failed mountain range. The Medicine Bow Mountains and the Sierra Madre stood off to the south, and while they and other ranges were rising this one had tried, too, but had succeeded merely in warping the flat land. The tilt of the strata was steeper than the road. Therefore, as we moved from cut to cut we were descending in time, downsection, each successive layer stratigraphically lower and older than the one before. Had this been a May morning a hundred million years ago, in Cretaceous time, we would have been many fathoms underwater, in a broad arm of the sea, which covered the continental platform—reached across the North American craton, the Stable Interior Craton—

from the Gulf of Mexico to the Arctic Ocean. The North Platte, scratching out the present landscape, had worked itself down into some dark shales that had been black muds in the organic richness of that epicratonic sea. The salt water rose and fell, spread and receded through time—in Love's words, "advanced westward and then retreated, then advanced and retreated over and over again, leaving thick sequences of intertonguing sandstone and shale"—repeatedly exposing fresh coastal plains, and as surely flooding them once more. In what has become dry mountain country, vegetation rioted in coastal swamps. They would have been like the Florida Everglades, the peat fens of East Anglia, or borders of the Java Sea, which stand just as temporarily, and after they are flooded by a rising ocean may be buried under sand and mud, and reported to the future as coal. There were seams of coal in the roadcuts, under the layers of sandstone and shale. The Cretaceous swamps were particularly abundant in this part of Wyoming. A hundred million years later, the Union Pacific Railroad would choose this right-of-way so it could fuel itself with the coal.

In cyclic rhythm with the other rock was limestone. Here and again, the highway was running on this soft impure limestone. It was sea-bottom lime, from dissolved or fragmented shells, which had lithified at least ten thousand feet lower than it is now. Woody asters were in bloom in the median, and blooming, too, by the side of the road, prospering on the lime. Love pointed them out with an edge in his voice. He said they were not Wyoming plants. They had come into Wyoming with trail herds of cattle and sheep, and later in trucks and railroad cars bringing hay from hundreds of miles to the south; and disastrously they had the ability—actually, a need—to draw selenium from the rock below. Selenium, which in concentration is toxic to people and animals, is given to the wind in some volcanic ash. A hundred million years ago, stratovolcanoes stood in Idaho, and they sent up ash that fell out eastward in the sea. The selenium went into the lime muds, and now these alien asters were drawing it out of the limestone and spreading the poison across the surface world, as few other plants can do. Most plants ignore selenium. Woody asters and a few others require selenium in order to germinate. After they take it up from the rock, they convert it into a form that nearly all plants will, in turn, take up, too. Selenium-contaminated plants are eaten by sheep and cattle, which are served to people as chops and burgers. Concentrated selenium destroys an enzyme that transmits messages from brain to muscles. "Cattle and sheep get the blind staggers," Love went on. "People are also affected. They get dishrag heart. The liver is damaged, and the kidneys. Selenium causes sterility. Worse, it causes birth defects. It's a cumulative poison, like lead or arsenic. It's one of the ingredients of nerve gas."

He gestured left and right. "These were prize salt-sage flats for sheep-grazing once. They're now poisoned and dangerous. A bad selenium area stinks like rotten garlic. On a warmer day, you could smell it. Fifty years ago, one of my first jobs was to look for selenium-converting plants up the Gros Ventre River. We camped there for a week, hunted for them day after day, and found a handful. Now, in the same place, they're thicker than fleas on a dog. They can't cross non-seleniferous barriers, except with the help of human beings. In the Rocky Mountains generally, millions of acres have been converted. People sometimes think neighbors have poisoned their pastures."

Ten miles beyond the North Platte, a flat-topped ridge formed the horizon before us—a tough sandstone, disintegrating at a lower rate than surrounding shale. The interstate, encountering this obstacle, had dealt with it with dynamite, opening up what highway engineers call a benched throughcut and geologists finding such a thing in nature call a wind gap. When we reached it, we stopped, got out, and put our noses on the outcrop, for this high multitiered exposure was Frontier sandstone, and Love referred to it as "a published roadcut," studied to the last grain by paleontologists and stratigraphers. The reason for so much attention was not readily apparent in the gray and somewhat gloomy, sooty-looking rock, antiqued with fossil burrows. Nonetheless, it seemed to excite Love— as he picked at it with his hammer—at least as much as the woody asters had repelled him. The rock had been submarine sand, not far offshore. "Frontier is one of the great oil sands in the Rocky Mountain region," he said. At five, ten, twenty thousand feet, wildcat after wildcat had found handsome pay in this celebrated host formation, and here it was at the surface, fresh, unweathered, presenting clues to its wealth. Oil almost always moves from one place to another, from source rock to host rock—from, for example, a buried lagoon, where it forms, to a fossil beach, whose permeable sands it fills. Petroleum is the transmuted remains of marine algae and other organic debris, which must first be buried in a manner that prevents oxidation—here in Cretaceous Wyoming by the transgressing muds of those shifting seas. Later, as an accident of sedimentation and tectonics, the organic remains must be held in a certain narrow range of temperature (not much above and not much below the temperature of boiling water) for at least a million years. That temperature range is known as the petroleum window. After oil forms, it is vulnerable to destruction by increased heat—in the earth as in the engine of a car. The oldest oil that has been recovered in commercial quantity is Ordovician in age (about four hundred and fifty million years). For about one human life span, geologists have had the ability to discern where, in the subsurface, oil should be. A large percentage of all the oil on earth has been burned up in fifty years. Around 1975, the quantity being discovered was surpassed, apparently forever, by the quantity being burned. Love remarked that half a billion barrels of oil had been found in the Frontier sandstone in one field alone. With reverence, I collected a wormy chunk.

Less than a mile up the road, we stopped again—at a low, flaky roadcut of Mowry shale. Progressing thus across Wyoming with David Love struck me as being analogous to walking up and down outside a theatre in the company of David Garrick. The classic plays—Teton, Beartooth, Wind River—were not out here on the street, but meanwhile these roadcuts were like posters, advertising the dramatic events, suggesting their narratives, fabrics, and structures. This Mowry shale had been organic mud of the Cretaceous seafloor, wherein the oil of the Frontier could have formed. It

was a shale so black it all but smelled of low tide. In it,
like mica, were millions of fish scales. It was interlayered
with bentonite, which is a rock so soft it is actually
plastic—pliable and porous, color of cream, sometimes
the color of chocolate. Bentonite is volcanic tuff—de-
composed, devitrified. So much volcanic debris has
settled on Wyoming that bentonite is widespread and,
in many places, more than ten feet thick. To some extent, it covers every basin. Also known as mineral soap,
it has the bizarre ability to adsorb water up to fifteen
times its own volume, and when this happens it offers
to a tire about as much resistance as soft butter. Wet,
swollen bentonite soil is known as gumbo. We were
crossing badlands of the Bighorn Basin one time when
a light shower fell, and the surface of the road changed
in moments from dust to colloidal suspension. The
wheels began to skid as if they were climbing ice.
Four-wheel drive was no help. Many a geologist has
walked out forty miles from a vehicle shipwrecked in
gumbo. Bentonite is mined in Wyoming and sold to the
rest of the world. Blessed is the land that can sell its
mud. Bentonite is used in adhesives, automobile polish,
detergent, and paint. It is in the drilling "mud" of oil
rigs, sent down the pipe and through apertures in the
bit to carry rock chips to the surface. It sticks to the
walls of the drill hole and keeps out unwanted water.
It is used to line irrigation ditches and reservoirs, and
in facial makeup. Indians drove buffalo into swamps
full of bentonite. It is an ingredient of insecticides, in-
sect repellents, and toothpaste. It is used to clarify beer.

If Love had ever tried bentonite to repair his air
mattress, he did not mention it. He did remark that

[11]

when there was rain in the Wind River Basin on the ranch where he grew up—an event that happened about as often as a birthday—wagons were stopped in their tracks. Much of Wyoming's bentonite is Cretaceous in age and consistent in composition. Since it lies on every side of the mountain ranges, it seems not so much to imply as to certify that when it was so broadly deposited the mountains were not there. The Cretaceous is not far back in the history of the world. It's in the last three per cent of time.

Love walked back to the Bronco with a look on his face that suggested a man who had long since had his last beer. He said he was hungry. He said, "My belly thinks my throat's been cut." Over the next rise was Rawlins, spread across the Union Pacific.

On October 20, 1905, the two-horse stage left Rawlins soon after dawn—not a lot of time for stretching out the comforts of the wonderful Ferris Hotel. Eggs were packed under the seats, also grapes and oysters. There were so many boxes and mailbags that they were piled up beside the driver. On the waybill, the passengers were given exactly the same status as the oysters and the grapes. The young woman from Wellesley, running her eye down the list of merchandise, encountered her own name: Miss Ethel Waxham.

The passenger compartment had a canvas roof, and canvas curtains at the front and sides.

The driver, Bill Collins, a young fellow with a four days beard, untied the bow-knot of the reins around the wheel, and swung up on the seat, where he ensconced himself with one leg over the mail bags as high as his head and one arm over the back of his seat, putting up the curtain between. "Kind o' lonesome out here," he gave as his excuse.

There were two passengers. The other's name was Alice Amoss Welty, and she was the postmistress of Dubois, two hundred miles northwest. Her post office was unique, in that it was farther from a railroad than any other in the United States; but this did not inconvenience the style of Mrs. Welty. Not for her some false-fronted dress shop with a name like Tinnie Mercantile. She bought her clothes by mail from B. Altman & Co., Manhattan. Mrs. Welty was of upper middle age, and —"bless her white hairs"—her gossip range appeared to cover every living soul within thirty thousand square miles, an interesting handful of people. The remark about the white hairs—like the description of Bill Collins and the estimated radius of Mrs. Welty's gossip —is from the unpublished journal of Ethel Waxham. The stage moved through town past houses built of railroad ties, past sheepfolds, past the cemetery and the state penitentiary, and was soon in the dust of open country, rounding a couple of hills before assuming a northwesterly course. There were limestone outcrops in the sides of the hills, and small ancient quarries at the base of the limestone. Indians had begun the quarries, removing an iron oxide—three hundred and fifty million years old—that made fierce and lasting warpaint. More recently, it had been used on Union Pacific railroad cars and, around 1880, on the Brooklyn Bridge. The hills above were the modest high points in a landscape that lacked exceptional relief. Here in the middle of the Rocky Mountains were no mountains worthy of the name.

Mountains were far away ahead of us, a range rising from the plains and sinking down again into them. Almost all the first day they were in sight.

As Wyoming ranges go, these distant summits were unprepossessing ridges, with altitudes of nine and ten thousand feet. In one sentence, though, Miss Waxham had intuitively written their geologic history, for they had indeed come out of the plains, and into the plains had in various ways returned.

Among rolling sweeps of prairie . . . we met two sheep herders with thousands of sheep each. "See them talking to their dogs," said the driver. They raised their arms and made strange gestures, while the dogs, at the opposite sides of the flock, stood on their hind legs to watch for orders.

In Wyoming in 1905, three million sheep competed for range grass with eight hundred thousand cattle. Big winter winds, squeezed and therefore racing fast between the high ground and the stratosphere, blew the snow off the grass and favored the sheep. They were hardier, and their wool contended with the temperature and the velocity of the wind. Winter wind. There was a saying among homesteaders in Wyoming: "If summer falls on a weekend, let's have a picnic."

Twelve miles from Rawlins, the horses were changed at Bell Spring, where, in a kind of topographical staircase—consisting of the protruding edges of sediments that dipped away to the east—the whole of the Mesozoic era rose to view: the top step Cretaceous, the next Jurassic, at the bottom a low red Triassic bluff, against which was clustered a compound of buildings roofed with cool red mud. Miss Waxham had no idea then that she was looking at a hundred and seventy-five million years, let alone *which* hundred and seventy-five million years. She had no idea that those sediments had broken off just here, and that the other side of the break, two and three thousand feet below, contained prolific traps of gas and oil. Actually, no one knew that. Discovery was twenty years away.

The stage rolled onto Separation Flats—altitude seven thousand feet—still pursuing the chimeric mountains. One of them, she learned, was called Whiskey Peak. Collins looked around from the driver's seat and said a passenger had once asked him the name of the mountain, "and I told him that it was in this coach where I could put my hand on it—but he could not guess." In the far distance also appeared a "white speck" —a roadhouse—which they watched impatiently for hours.

It did not look larger when we reached it. . . . Mrs. Welty and I hurried in to get warm, for we were chilled

[15]

through. Outside, hung from the roof, was half a carcass of a steer. . . . In a cluttered kitchen, a fat forlorn silent woman served us wearily with a plentiful but plain meal, and sat with her arms folded watching us eat. . . . We ate our baked potatoes and giant biscuits, onions and carrots and canned-apple pie in half silence, glad to be through. The stage horses were changed and we started on toward Lost Soldier.

Lost Soldier was another sixteen miles and thus would take three hours. Already, Mrs. Welty was talking about the Hog Back, more than twenty hours up the road—a steep descent from a high divide, where Wyoming's storied winds had helped many a stagecoach get to the bottom in seconds. Wreckage was strewn all over the ground there, among the bones of horses. A driver had been known to chain a coach to a tree to keep the coach from blowing away. Like the sails of boats and ships, the canvas sides of stagecoaches were often furled as they approached the Hog Back, to let the wind blow through. No one relied on brakes.

Always, going down, the wheels are rough-locked by a chain so that they slip along instead of turning. . . . A freight team went over the side a little while ago about Thanksgiving time. The load was partly supplies for Thanksgiving dinner, turkeys, oysters, fruit, etc. The driver called to the team behind for help. When it came, he was calmly seated on a stump peeling an orange while the wagon and debris were scattered below.

Oil would be discovered under Lost Soldier in 1916. It would yield the highest recovery per acre of any oil field that has ever been discovered in the Rocky Mountains. From level to level in a drill hole there—a hole about a mile deep—oil could be found in an amazing spectrum of host rocks: in the Cambrian Flathead sandstone, in the Mississippian Madison limestone, in the Tensleep sands of Pennsylvanian time. Oil was in the Chugwater (red sands of the Triassic), and in the Morrison, Sundance, Nugget (celebrated formations of the Jurassic), and, of course, in the Cretaceous Frontier. A well at Lost Soldier was like grafted ornamental citrus —oranges, lemons, tangerines, grapefruit, all on a single tree. The discoverer of the oil-bearing structure was a young geologist from Princeton University, who not only found the structure but also helped to place the term "sheepherder anticline" in the geologic lexicon. A sheepherder anticline is one that is particularly obvious, one that could be mapped by a Princeton geologist dressed as a shepherd and moving around with a flock of sheep—which is how he avoided attention as he studied the rock of Lost Soldier.

We rattled into the place at last, and were glad to get in to the fire to warm ourselves while the driver changed the load from one coach to another. With every change of drivers the coach is changed, making each man responsible for repairs on his own coach. The Kirks keep Lost Soldier. Mrs. Kirk is a short stocky figureless woman with untidy hair. She furnished me with an old soldier's overcoat to wear during the night to come. . . . Before long, we were started again, with Peggy Dougherty for driver. He is tall and grizzled. They say that when he goes to dances they make him take the spike out of the bottom of his wooden leg.

There were four horses now—"a wicked little team"—and immediately they kicked over the traces, tried to run away, became tangled like sled dogs twisting in harness, and set Peggy Dougherty to swearing.

Ye gods, how he could swear.

Mrs. Welty diverted him with questions about travellers marooned in snowdrifts. Mrs. Welty was aware that Mr. Dougherty—who was missing six fingers, one leg, and half of his remaining foot—was an authority on this topic. In 1883, a blizzard had overtaken him and his one passenger, a young woman comparable in age to Miss Waxham. When the snow became so deep that the coach ceased to move, he unhitched a horse. Already stiff with frostbite, he hung on to the harness while the animal hauled him through drifts. The horse dragged Dougherty for hours, until he finally lost his grip and let go, having nearly reached a stage-line station. Into the wind, he shouted successfully for help. When rescuers reached the stagecoach, the passenger was dead.

Dougherty remarked to Mrs. Welty that winters lately had not been so severe.

"No," she agreed. "And we haven't had a blizzard this summer."

The sun set, and the stars rose, and the cold grew more intense. . . . About half past nine, we reached the supper station, stiff with cold.

This was Rongis, a community of a few dozen people just south of Crooks Gap. "Rongis" was an ananym —so named by an employee of the stagecoach company whose own name was Eli Signor. Lost Soldier, Rongis— such names are absent now among the Zip Codes of Wyoming, but the ruins of the stations remain.

Supper was soon ready, a canned supper, with the usual dried-apple pie and monstrous biscuits and black coffee. About ten we started out again, with a new relay of horses. More wrapped up than ever, we sat close to each other to keep warm, and leaned against the sacks of mail behind us.

The night before, at Rongis, the temperature had gone to zero. As the stage moved into Crooks Gap, the bright starlight fell on fields of giant boulders black-and-silver in relief. Some were as large as houses. In time, it would be determined that they had come down off high mountains farther north that were no longer high —mountains that had somehow sunk into the plain. Meanwhile, anyone connecting the boulders to their source bedrock might wonder how they had made their way uphill. The big boulders were granite, and smaller ones among them—recognized by no one then—were jade: float boulders of gem jade, nephrite jade (green as emerald), rounded in streambeds and polished by weather. As she watched them in the moonlight from the stage, they must have seemed just rubble on the ground. There was uranium in Crooks Gap in great quantity—in pods and lenses for a thousand feet up either side. It would be discovered in 1955. There was petroleum under Crooks Gap, too. The year of discovery would be 1925. Crooks Creek flowed through Crooks Gap—straight through the highlands, from one side to the other. Above the gap was Crooks Mountain. Miss Waxham might well have wondered who the eponymous crook was. The possibilities in that country were bewilderingly numerous, but the honor belonged to Brigadier General George Crook, West Point '52, known among the Indians as the Gray Fox. General Crook, commander of the Department of the Platte, was at least a century ahead of his time in the integrity with which he dealt with aboriginal people, and deserved having his name writ in land if for no other reason than his reply when someone asked him if the campaigns of the Indian wars were difficult work. He said, "Yes, they are hard. But the hardest thing is to go out and fight against those who you know are in the right."

I watched for hours the shadow of the suitcase handle against the canvas to see the moon's change of position. The hours dragged by, and the cold grew worse. . . . Between three and four we reached Myersville.

They had come to the Sweetwater River, which they forded, with still another driver, who had a remarkably delicate cast of tongue. "Oh, good gracious!" he shouted at the team.

The driver had been on the road only once, did not know his horses, and had no whip. The Hog Back was ahead. . . . There was no more sleep for us then, not an eye wink.

The Hog Back was a knife-edged spur plunging off the Beaver Divide, which separates waters that flow east into the Platte from waters that flow north into the Wind, Bighorn, and Yellowstone Rivers. The Hog Back was Frontier sandstone and Mowry shale, which had accumulated flat in the Cretaceous sea, and here, in subsequent time, had been bent upward sharply to make the jagged edge the travellers descended. Its shales were slick with bentonitic gumbo.

At the top of Beaver Mountain we saw the Wind River Range stretching white in the distance. The driver rough-locked the back wheels and we started down. It was a scramble for the horses to keep out of the way. There were sudden turns in the road and furrows cut by the freight wagons that almost threw the careening stage on its side. One of the horses fell, but was dragged along by the others until it finally regained its feet. We finally reached a place where the slope was less steep, the rough lock was taken off, and the driver began again to try to make time down the hills. The little leaders ran like rats and the heavier wheelers were carried along while the coach swung from side to side in the gullies.

Twenty-six hours out of Rawlins, the stage reached Hailey. Breakfast was waiting, and in Miss Waxham's opinion could have gone on waiting—"the same monstrous biscuits and black coffee." A rancher named Gardiner Mills arrived—"short, dark, of caustic speech" —and handed her a big fur overcoat to top her own and keep her warm in his springy buckboard. He had come to take the new schoolmarm the remaining ten miles to his Red Bluff Ranch, and into the afternoon they travelled northwest under six-hundred-foot walls of rose, vermilion, brick, and carmine—red Triassic rock. Near a big spring under the red bluff were the low buildings of the ranch.

The corral and bunkhouse, grain and milk house were log structures off to one side. When we drove up to the gate, and two little narrow-chested large-pompadoured girls came out the walk to meet us, all my fears as to obstreperous pupils were at an end.

The "chiffonier" in her room was a stack of boxes covered with muslin curtains. There was "a washstand for private individual use." There was a mirror a foot square. On her walls were Sargent and Gainsborough prints, and pictures of Ethel Barrymore and Psyche.

In the western outskirts of Rawlins, David Love pulled over onto the shoulder of the interstate, the better to fix the scene, although his purpose in doing so was not at all apparent. Rawlins reposed among low hills and prairie flats, and nothing in its setting would ever lift the stock of Eastman Kodak. In those western outskirts, we may have been scarcely a mile from the county courthouse, but we were very much back on the range—a dispassionate world of bare rock, brown grass, drab green patches of greasewood, and scattered colonies of sage. The interstate had lithified in 1965 as white concrete but was now dark with the remains of ocean algae, cremated and sprayed on the road. To the south were badlands—gullies and gulches, erosional debris. To the north were some ridgelines that ended sharply, like breaking waves, but the Rawlins Uplift had miserably fallen short in its bid to be counted among the Rocky Mountains. So why was David Love, who had the geologic map of Wyoming in his head, stopping here?

The rock that outcropped around Rawlins, he said, contained a greater spread of time than any other suite of exposed rocks along Interstate 80 between New York and San Francisco. We were looking at many moments in well over half the existence of the earth, and we were seeing—as it happened—a good deal more time than one sees in the walls of the Grand Canyon, where the clock stops at the rimrock, aged two hundred and fifty million years. The rock before us here at Rawlins reached back into the Archean eon and up to the Miocene epoch. Any spendthrift with a camera could aim it into that scene and—in a two-hundred-and-fiftieth of a second at f/16—capture twenty-six hundred million years. The most arresting thing in the picture, however, would be Rawlins' municipal standpipe—that white, squat water-storage tank over there on the hill.

The hill, though, was Archean granite and Cambrian sandstone and Mississippian limestone. If you could have taken pictures when *they* were forming, the collection would be something to see. There would be a deep and uncontinented ocean sluggish with amorphous scums (above cooling invisible magmas). Ther would be a risen continent reaching its coast, with rivers running over bare rock past not so much as a lichen. There would be rich-red soil on a broad lowland plain resembling Alabama (but near the equator). There would be clear, warm shelf seas.

There would also be a picture of dry hot dunes, all of them facing the morning sun—the rising Miocene sun. Other—and much older—dunes would settle a great question, for it is impossible to tell now whether they were just under or just out of water. They covered all of Wyoming and a great deal more, and may have been very much like the Empty Quarter of Arabia: the Tensleep-Casper-Fountain Pennsylvanian sands. There would be a picture, too, of a meandering stream, with overbank deposits, natural levees, cycads growing by the stream. Footprints the size of washtubs. A head above the trees. In the background, swamp tussocks by the shore of an oxbow lake. What was left of that picture was the Morrison formation—the Jurassic landscape of particularly dramatic dinosaurs—outcropping just up the road. There would be various views of the great Cretaceous seaway, with its plesiosaurs, its giant turtles, its crocodiles. There would be a picture from the Paleocene of a humid subtropical swamp, and a picture from the Eocene of gravel bars in a fast river running off a mountain onto lush subtropical plains, where puppy-size horses were hiding for their lives.

Such pictures, made in this place, could form a tall stack—scene after scene, no two of them alike. Taken together, of course—set one above another, in order—they would be the rock column for this part of Wyoming. They would correlate with what one would see in the well log of a deep-drilling rig. There would be hiatuses, to be sure. In the rock column, anywhere, more time is missing than is there; so much has been eroded away. Besides, the rock in the column is more apt to commemorate a moment—an eruption, a flood, a fallen drop of rain—than it is to report a millennium. Like a news broadcast, it is more often a montage of disasters than a cumulative record of time.

I asked Love why so much of the earth's history happened to be here on the surface in this nondescript part of the state.

He said, "It just came up in the soup. Why it is out here all by itself is a matter of fierce debate." The Rawlins Uplift had not accomplished nothing.

The Precambrian granite on the ridge was from the end of the two billion years that are covered in the first nine verses of Genesis, when there finally was continent cool enough and strong enough to stand above the sea: when God said, "Let the waters under the heaven be gathered together unto one place, and let the dry land appear"—at the end of the Archean eon.

The granite dipped below us, and close to the interstate the Union Pacific had been blasted through some sandstone that rested on, and was derived from, the granite—littoral sands of Cambrian time, when the American west coast was at Rawlins. Between this Flathead sand and the Madison limestone above it, lying here and there in pockets in an unconformity of a hundred and seventy million years, was the rich-red soil of the Paleozoic plain. A streak of it showed in a low hillside even closer to us than the railroad cut, so we walked over to collect some and put it in a bag. As rock it was so incompetent that it could easily be crushed to powder—a beautiful rose-brick powder with the texture of cocoa. It had been known in the paint business as Rawlins Red, and in the warpaint business as effective medicine, this paleosol (fossil soil) three hundred and fifty million years old. As we returned to the road, a couple of Consolidated Freightways three-unit twenty-six-wheel tractor-trailers went by, imitating thunder. Love said, "First we had the Conestoga, then the big freight wagons with twelve to sixteen oxen. Now we have those things."

The spread of time at Rawlins, like the rock column in a great many places in Wyoming, was so impressively detailed that it seemed to suggest that Wyoming, in its one-thirty-seventh of the United States, contains a disproportionate percentage of American geology. Geologists tend to have been strongly influenced by the rocks among which they grew up. The branch of the science called structural geology, for example, has traditionally been dominated by Swiss, who spend their youth hiking and schussing in a national textbook of structure. When a multinational oil company held a conference in Houston that brought together structural geologists from posts all over the world, the coffee breaks were in Schweizerdeutsch. The wizards of sedimentology tend to be Dutch, as one would expect of a people who have figured out a way to borrow against unrecorded deposits. Cincinnati has produced an amazingly long list of American paleontologists—Cincinnati, with profuse exceptional fossils in its Ordovician hills. Houston—the capital city of the oil geologist—is a hundred and fifty miles from the first place where you can hit a hammer on a rock. Houston geologists come from somewhere else.

Geologists who have grown up on shield rock—Precambrian craton—tend to be interested in copper, diamonds, iron, and gold. Most of the world's large metal deposits are Precambrian. Diamonds, after starting upward from the mantle, seem to need the thickness of a craton to survive their journey to the surface world. Geologists who grow up on shield rock also tend to be impressed by the fact that it has been around since the earliest development of life, and to imagine a progression through time in which the uniformitarian recycling of the earth's materials is a mere subplot in a dramatic narrative that begins with dark scums in motion on an otherwise watery globe and evolves through various continental configurations toward the scenery of the earth today.

Geologists who grow up in California start out with strange complex structures, highly deformed rock—mélanges and turbidites that seem less in need of a G. K. Gilbert than of an Alfred Adler or a Carl Jung. Shell, in its rosters, used to put an asterisk beside the names of geologists from California. The asterisk meant that while they were in, say, Texas they might be quite useful among the Gulf Coast turbidites of the Hackberry Embayment, but assign them with caution almost anywhere else. A former Shell geologist (not David Love) once said to me, "The asterisk also meant 'Ship them back to California when they're done.' Shell considered them a separate race."

A geologist who grew up in Wyoming would have something of everything above—with the probable exception of the asterisk. A geologist who grew up in Wyoming could not ignore economic geology, could not ignore vertebrate paleontology, could not ignore the narrative details in any chapter of time (every period in the history of the world was represented in Wyoming). Wyoming geology would above all tend to produce a generalist, with an eye that had seen a lot of rocks, and a four-dimensional gift for fitting them together and arriving at the substance of their story— a scenarist and lithographer of what geologists like to call the Big Picture.

Wyoming, at first glance, would appear to be an arbitrary segment of the country. Wyoming and Colorado are the only states whose borders consist of four straight lines. That could be looked upon as an affront to nature, an utterly political conception, an ignoring of the outlines of physiographic worlds, in disregard of rivers and divides. Rivers and divides, however, are in some ways unworthy as boundaries, which are meant to imply a durability that is belied by the function of rivers and divides. They move, they change, and they go away. Rivers, almost by definition, are young. The oldest river in the United States is called the New River. It has existed (in North Carolina, Virginia, and West Virginia) for a little more than one and a half per cent of the history of the world. In epochs and eras before there ever was a Colorado River, the formations of the Grand Canyon were crossed and crisscrossed, scoured and dissolved, deposited and moved by innumerable rivers. The Colorado River, which has only recently appeared on earth, has excavated the Grand Canyon in very little time. From its beginning, human beings could have watched the Grand Canyon being made. The Green River has cut down through the Uinta Mountains in the last few million years, the Wind River through the Owl Creek Mountains, the Laramie River through the Laramie Range. The mountains themselves came up and moved. Several thousand feet of basin fill has recently disappeared. As the rock around Rawlins amply shows, the face of the country has frequently changed. Wyoming suggests with emphasis the page-one principle of reading in rock the record of the earth: Surface appearances are only that; topography grows, shrinks, compresses, spreads, disintegrates, and disappears; every scene is temporary, and is composed of fragments from other scenes. Four straight lines—like a plug cut in the side of a watermelon—should do as well as any to frame Wyoming and its former worlds.

A geologist who grew up in Wyoming has grown up among mountains that in terms of plate-tectonic theory are the least explainable in the world. A geologist who grew up in Wyoming—with its volcanic activity, its mountains eroding, and its basins receiving sediment— would inherently comprehend the cycles of the earth: geology repeating itself as people watch. G. K. Gilbert, the first Chief Geologist of the United States Geological Survey, once remarked that it is "the natural and legitimate ambition of a properly constituted geologist to see a glacier, witness an eruption and feel an earthquake." A geologist could do all that as a child in Wyoming, and not have to look far for more.

 Miss Waxham's school was a log cabin on Twin
Creek near the mouth of Skull Gulch, a mile from the
Mills ranch. Students came from much greater distances, even through deep snow. Many mornings, ink
was frozen in the inkwells, and the day began with
ink-thawing, followed by reading, spelling, chemistry,
and civil government. Sometimes snow blew through
the walls, forming drifts in the schoolroom. Water was
carried from the creek—drawn from a hole that was
chopped in the ice. If the creek was frozen to the
bottom, the students melted snow. Their school was
fourteen by sixteen feet—smaller than a bathroom at
Wellesley. The door was perforated with bullet holes
from "some passerby's six-shooter." Over the ceiling
poles were old gunnysacks and overalls, to prevent the
sod roof from shedding sediment on the students. Often,
however, the air sparkled with descending dust, struck
by sunlight coming in through the windows, which were
all in the south wall. There was a table and chair for
Miss Waxham, and eight desks for her pupils. Miss
Waxham's job was to deliver a hundred per cent of the
formal education available in District Eleven, Fremont
County, Wyoming.

The first fifteen minutes or half hour are given to reading "Uncle Tom's Cabin" or "Kidnapped," while we all sit

about the stove to keep warm. Usually in the middle of a reading the sound of a horse galloping down the frozen road distracts the attention of the boys, until a few moments later six-foot George opens the door, a sack of oats in one hand, his lunch tied up in a dish rag in the other. Cold from his five-mile ride, he sits down on the floor by the stove, unbuckles his spurs, pulls off his leather chaps, drops his hat, unwinds two or three red handkerchiefs from about his neck and ears, takes off one or two coats, according to the temperature, unbuttons his vest and straightens his leather cuffs. At last he is ready for business.

· · ·

Sandford is the largest scholar, six feet, big, slow in the school room, careful of every move of his big hands and feet. His voice is subdued and full of awe as he calls me "ma'am." Outside while we play chickens he is another person—there is room for his bigness. Next largest of the boys is Otto Schlicting, thin and dark, a strange combination of shrewdness and stupidity. His problems always prove, whether they are right or not! He is a boaster, too, tries to make a big impression. But there is something very attractive about him. I was showing his little sister how to add and subtract by making little lines and adding or crossing off others. Later I found on the back of Otto's papers hundreds and hundreds of little lines—trying to add that way as far as a hundred evidently. He is nearly fifteen and studying division. . . . Arithmetic is the family failing. "How many eights in ninety-six?" I ask him. He thinks for a long time. Finally he says— with such a winsome smile that I wish with all my heart it were true—"Two." "What feeds the cells in your body?" I ask him. He thinks. He says, "I guess it's vinegar." He has no idea of form. His maps of North America on the board are all like turnips.

Students' ages ranged through one and two digits, and their intelligence even more widely. When Miss Waxham called upon Emmons Schlicting, asking, "Where does digestion take place?," Emmons answered, "In the Erie Canal." She developed a special interest in George Ehler, whose life at home was troubled.

He is only thirteen, but taller than Sandford, and fair and handsome. I should like to get him away from his family —kidnap him. To think that it was he who tried to kill his father! His face is good as can be.

At lunchtime, over beans, everyone traded the news of the country, news of whatever might have stirred in seven thousand square miles: a buffalo wolf trapped by Old Hanley; missing horses and cattle, brand by brand; the sheepherder most recently lost in a storm. If you went up Skull Gulch, behind the school, and climbed to the high ground beyond, you could see seventy, eighty, a hundred miles. You "could see the faint outlines of Crowheart Butte, against the Wind River Range." There was a Wyoming-history lesson in the naming of Crowheart Butte, which rises a thousand feet above the surrounding landscape and is capped with flat sandstone. To this day, there are tepee rings on Crowheart Butte. One of the more arresting sights in remote parts of the West are rings of stones that once resisted the wind and now recall what blew away. The Crows liked the hunting country in the area of the butte, and so did the Shoshonis. The two tribes fought, and lost a lot of blood, over this ground. Eventually, the chief of the Shoshonis said, in effect, to the chief of the

Crows: this is pointless; I will fight you, one against one; the hunting ground goes to the winner. The chief of the Shoshonis was the great Washakie, whose name rests in six places on the map of Wyoming, including a mountain range and a county. Washakie was at least fifty, but fit. The Crow would have been wise to demur. Washakie destroyed him in the hand-to-hand combat, then cut out his heart and ate it.

Despite her relative disadvantages as a newcomer, an outlander, and an educational ingénue, Miss Waxham was a quick study. Insight was her long suit, and in no time she understood Wyoming. For example, an entry in her journal says of George Ehler's father, "He came to the country with one mare. The first summer, she had six colts! She must have had calves, too, by the way the Ehlers' cattle increased." These remarks were dated October 22, 1905—the day after her stagecoach arrived. In months that followed, she sketched her neighbors (the word applied over many tens of miles). "By the door was Mrs. Frink, about 18, with Frink junior, a large husky baby. Ida Franklin, Mrs. Frink's sister and almost her double, was beside her, frivolous even in her silence." There was the story of Dirty Bill Collins, who had died as a result of taking a bath. And she fondly recorded Mrs. Mills' description of the libertine Guy Signor: "He has a cabbage heart with a leaf for every girl." She noted that the nearest barber had learned his trade shearing sheep, and a blacksmith doubled as dentist. Old Pelon, a French Canadian, impressed her, because he had refused to ask for money from the government after Indians killed his brother. "Him better dead," said Old Pelon. Old Pelon was fond of the masculine objective pronoun. Miss Waxham wrote, "Pelon used to have a wife, whom he spoke of always as 'him.'" Miss Waxham herself became a character in this tableau. People sometimes called her the White-Haired Kid.

"There's many a person I should be glad to meet," read an early entry in her journal. She wanted to meet Indian Dick, who had been raised by Indians and had no idea who he was—probably the orphan of emigrants the Indians killed. She wanted to meet "the woman called Sour Dough; Three Fingered Bill, or Suffering Jim; Sam Omera, Reub Roe. . . ." (Reub Roe held up wagons and stagecoaches looking for members of the Royal Family.) Meanwhile, there was one flockmaster and itinerant cowboy who seemed more than pleased to meet her.

In the first reference to him in her journal she calls him "Mr. Love—Johnny Love." His place was sixty miles away, and he had a good many sheep and cattle to look after, but somehow he managed to be right there when the new young schoolmarm arrived. In the days, weeks, and months that followed, he showed a pronounced tendency to reappear. He came, generally, in the dead of night, unexpected. Quietly he slipped into the corral, fed and watered his horse, slept in the bunkhouse, and was there at the table for breakfast in the morning—this dark-haired, blue-eyed, handsome man with a woolly Midlothian accent.

Mr. Love is a Scotchman about thirty-five years old. At first sight he made me think of a hired man, as he lounged stiffly on the couch, in overalls, his feet covered with enormous red and black striped stockings that reached to his knees, and were edged with blue around the top. He seemed to wear them instead of house shoes. His face was kindly, with shrewd blue twinkling eyes. A moustache grew over his mouth, like willows bending over a brook. But his voice was most peculiar and characteristic. . . . A little Scotch dialect, a little slow drawl, a little nasal quality, a bit of falsetto once in a while, and a tone as if he were speaking out of doors. There is a kind of twinkle in his voice as well as his eyes, and he is full of quaint turns of speech, and unusual expressions.

Mr. Love travelled eleven hours on these journeys, each way. He did not suffer from the tedium, in part because he frequently rode in a little buggy and, after telling his horses his destination, would lie on the seat and sleep. He may have been from Edinburgh, but he had adapted to the range as much as anyone from anywhere. He had slept out, in one stretch, under no shelter for seven years. On horseback, he was fit for his best horses: he had stamina for long distances at sustained high speed. When he used a gun, he hit what he was shooting at. In 1897, he had begun homesteading on Muskrat Creek, quite near the geographical center of Wyoming, and he had since proved up. One way and another, he had acquired a number of thousands of acres, but acreage was not what mattered most in a country of dry and open range. Water rights mattered most, and the area over which John Love controlled the water amounted to a thousand square miles—about one per cent of Wyoming. He had come into the country walking, in 1891, and now, in 1905, he had many horses, a couple of hundred cattle, and several thousand sheep. Miss Waxham, in her journal, called him a "mutton-aire."

He was a mirthful Scot—in abiding contrast to the more prevalent kind. He was a wicked mimic, a connoisseur of the absurd. If he seemed to know everyone in the high country, he knew even better the conditions it imposed. After one of her conversations with him, Miss Waxham wrote in her journal:

It is a cruel country as well as beautiful. Men seem here only on sufferance. After every severe storm we hear of people's being lost. Yesterday it was a sheep camp mover who was lost in the Red Desert. People had hunted for him for a week, and found no trace. Mr. Love—Johnny Love— told of a man who had just been lost up in his country, around the Muskrat. "Stranger?" asked Mr. Mills. "No; born and brought up here." "Old man?" "No; in the prime of life. Left Lost Cabin sober, too."

Mr. Love had been born near Portage, Wisconsin, on the farm of his uncle the environmentalist John Muir. The baby's mother died that day. His father, a Scottish physician who was also a professional photographer and lecturer on world travel, ended his travels and took his family home. The infant had three older sisters to look after him in Scotland. The doctor died when John was twelve. The sisters emigrated to Broken Bow, Nebraska, where in the eighteen-seventies and eighties they all proved up on homesteads. When John was in his middle teens, he joined them there, in time to experience the Blizzard of '88—a full week of blowing snow, with visibility so short that guide ropes led from house to barn.

He was expelled from the University of Nebraska for erecting a sign in a dean's flower bed, so he went to work as a cowboy, and soon began to think about moving farther west. When he had saved enough money, he bought matching black horses and a buggy, and set out for Wyoming. On his first night there, scarcely over the border, his horses drank from a poison spring and died. What he did next is probably the most encapsulating moment in his story. In Nebraska were three homes he could return to. He left the buggy beside the dead horses, abandoned almost every possession he had in the world, and walked on into Wyoming. He walked about two hundred miles. At Split Rock, on the Oregon Trail—near Crooks Gap, near Independence Rock—he signed on as a cowboy with the 71 Ranch. The year was 1891, and the State of Wyoming was ten months old.

Through the eighteen-nineties, there are various hiatuses in the résumé of John Love, but as cowboy and homesteader he very evidently prospered, and he also formed durable friendships—with Chief Washakie, for example, and with the stagecoach driver Peggy Dougherty, and with Robert LeRoy Parker and Harry Longabaugh (Butch Cassidy and the Sundance Kid). There came a day when Love could not contain his developed curiosity in the presence of the aging chief. He asked him what truth there was in the story of Crowheart Butte. Had Washakie really eaten his enemy's heart? The chief said, "Well, Johnny, when you're young and full of life you do strange things."

Robert LeRoy Parker was an occasional visitor at Love's homestead on Muskrat Creek, which was half-way between Hole-in-the-Wall and the Sweetwater River—that is, between Parker's hideout and his woman. Love's descendants sometimes stare bemusedly at a photograph discovered a few years ago in a cabin in Jackson Hole that had belonged to a member of the Wild Bunch. The photograph, made in the middle eighteen-nineties, shows eighteen men with Parker, who is wearing a dark business suit, a tie and a starchy white collar, a bowler hat. Two of the bunch are identified only by question marks. One of these is a jaunty man of middle height and strong frame, his hat at a rakish angle—a man with a kindly face, twinkling shrewd eyes, and a mustache growing over his mouth like willows bending over a brook. It may be doubtful whether John Love would have joined such a group, but when you are young and full of life you do strange things.

At Red Bluff Ranch, Mrs. Mills once twitted Mr. Love for being Scottish when other Scots were around and American in the presence of Americans. For a split second, Mr. Love thought this over before he said, "That leaves me eligible for the Presidency." Out of Mr. Love's buggy came a constant supply of delicacies and exotic gifts—including candy, nuts, apples—which he came by who knows where and liberally distributed to all. Miss Waxham began to look upon him as "a veritable Santa Claus"; and, predictably, at Christmas-time Santa appeared.

And the next day was Christmas. . . . Just before supper the joyful cry went up that Mr. Love was coming, and actually in time for dinner. He had broken his record and arrived by day!

A pitch pine had been set up indoors and its boughs painted with dissolved alum to simulate frost. Hanging from the branches were wooden balls covered with tobacco tinfoil. Flakes of mica were glued to paper stars. On Christmas, Mr. Mills and Mr. Love dressed in linen collars and what Miss Waxham called "fried shirts." When Miss Waxham turned to a package from home that she knew contained pajamas, she went into her bedroom to open it.

The following day, Miss Waxham was meant to go to something called Institute, in Lander—a convocation of Fremont County schoolteachers for lectures, instruction, and professional review. By phenomenal coincidence, Mr. Love announced that he had business in Lander, too.

It was decided that I should go with him. I rather dreaded it. . . . I confess I was somewhat afraid of him. . . . I was wrapped up in a coat of my own with Mrs. Mills' sealskin over it, muffler, fur hat, fur gloves, leggings, and overshoes. Then truly I was so bundled up that it was next to impossible to move. "Absolutely helpless," laughed Mr. Love.

Whatever business Mr. Love had in Lander did not in any way seem to press him. Miss Waxham stayed with Miss Davis, the county superintendent, and while other people came and went from the premises Mr. Love was inclined to remain.

Supper time came and Mr. Love remained. We had a miserable canned goods cold supper. Miss MacBride left, Mr. Love remained.

. . .

In the afternoon, Mr. Love called. It certainly was a surprise. I explained why Miss Davis was out, but he didn't seem to mind. I said that she would be back soon. He asked if I should not like to take a drive and see the suburbs. Of course I would. . . . We went for a long drive in the reservation, with a box of chocolates between us, and a merry gossip we had. . . . He was bemoaning the fact that there is no place for a man to spend the evening in Lander except in a saloon. "Come and toast marshmallows," I said, and he took it as a good suggestion.

When she went to church on Sunday, Love was there—John Santa Love, who had not been to church in ten years. After the service, it was time to leave Lander.

There had been snow falling since morning, and the road was barely visible. The light faded to a soft whiteness that hardly grew darker when the sun set and the pale outline of the moon showed through the snow. Everywhere was the soft enveloping snow shutting out all sounds and sights. The horses knew the way and travelled on steadily. Fortunately it was not cold, and the multitudinous rugs and robes with the new footwarmer beneath kept us warm and comfortable. More pleasant it was travelling through the storm than sitting at home by the fire and watching it outside. When the conversation ran low and we travelled on quietly, Mr. Love discovered bags of candy under the robes . . . and he fed us both, for I was worse than entangled in wraps and the long sleeves of Mrs. Mills' sealskin. The miles fell away behind us easily and quietly.

Even as those words were written, the editor and publishers of the *Shoshone Pathfinder*, in Lander, were completing a special issue urging young people to make their lives in central Wyoming. "We beg leave to extend to each and every one of you a most cordial welcome to come, remain, and help develop a country so rich in natural resources as to be beyond the computation of mortal man," wrote the publishers. It was a country "clothed in a mantle of the most nutritious grasses and sage brush browse." In its Wind River Mountains were "thousands of square miles of dense forests, which the foot of man has never invaded, and . . . as to the supply and quality of timber in this county it will meet the requirements of all demands for all time to come." Moreover, there was coal: "It has been said of our coal fields that the entire United States would be unable to exhaust them in a century. . . . It is in excess of the imagination to contemplate the vastness of this tremendous supply of fuel or what would ever transpire to exhaust it." And there was oil: "It is a recognized fact of long standing that the quantity of oil stored in the natural reservoirs of this county is so great that no estimate can be made." And there was gold. At the south end of the Wind Rivers, nearly five million dollars had come out of small mines with names like Hard Scrabble, Ground Hog, Hidden Hand, Mormon Crevice, Iron

Duke, Midget, Rustler, Cariboo, and Irish Jew. "None
of the mines have been exhausted, but merely sunk to
a depth where more and better machinery is required."
There was uranium, too, but as yet no compelling need
to find it, and as yet no geologist equal to the task.

As the winter continued, with its apparently inex-
haustible resources of biting wind and blinding snow,
temperatures now and again approached fifty below
zero. Miss Waxham developed such an advanced case of
cabin fever that she wrote in her journal, "My spirit has
a chair sore." Even when drifts were at their deepest,
though, Mr. Love somehow managed to get through.
"Much wrapped up" on one occasion, he rode "all the
way from Alkali Butte." On another, he spent an entire
day advancing his education at the Twin Creek school.

These attentions went on in much the same way for
five years. He pursued her to Colorado, and even to
Wisconsin. They were married on the twentieth of June,
1910, and drove in a sheep wagon to his ranch, in the
Wind River Basin. It was plain country with gently
swelling hills. Looking around from almost any one of
them, you could see eighty miles to the Wind River
Range, thirty to the Owl Creeks, twenty to the Rattlesnake Hills, fifteen to the Beaver Divide, and a hundred
into the Bighorns. No buildings were visible in any
direction. In this place, they would flourish. Here, too,
they would suffer calamitous loss. Here they would raise
three children—a pair of sons close in age, and, a dozen
years after them, a daughter. The county from time to
time would supply a schoolmarm, but basically the children would be educated by their mother. One would
become a petroleum chemist, another a design engineer

for the New Jersey Turnpike and the New York State Thruway, another the preeminent geologist of the Rocky Mountains.

⌒ Along the Nebraska-Wyoming line, in the region of the forty-first parallel, is a long lumpy break in the plains, called Pine Bluffs. It is rock of about the same age and story as Scotts Bluff, which is not far away. David Love—standing on top of Pine Bluffs—remarked that for a great many emigrants with their wagons and carts these had been the first breaks in the horizon west of Missouri. From the top of the bluffs, the emigrants had their first view of the front ranges of the Rockies, and the mountains gave them hope and courage. For our part, looking west from the same place, we could not see very far across the spring wildflowers into the swirling snow. The Laramie Range was directly ahead and the Never Summer Mountains off somewhere to the southwest—at ten and nearly thirteen thousand feet indeed a stirring sight, but not today. Love said a spring snowstorm was "sort of like a kiss—it's temporary, and it will go away." (That one stopped us for three days.) Meanwhile, there were large roadcuts to examine where Interstate 80 sliced through the bluffs, and scenes to envision that were veiled by more than snow.

The bluffs stand above the surrounding country because—like other mesas and buttes—they are all that is left of what was the surrounding country. The rock of Pine Bluffs is sedimentary (limestone, sandstone) and seems to lie flat, but in fact it tilts very slightly, and if its bedding planes are projected westward a hundred miles they describe the former landscape, rising about sixty feet a mile to touch the summits of mountains. "There was a continuous surface," Love said. "It came over the top of the Laramie Range and out here onto the High Plains. Pine Bluffs was part of that surface."

In earth history, that was not long ago. He said the best general date for it was ten million years—when the central Rocky Mountains, which had long since taken form as we know them, were buried up to their chins. Only the highest peaks remained uncovered, like nunataks protruding through continental ice, or scattered islands in a sea. The summits of the Wind River Range were hills above that Miocene plain. The highest of the Bighorns stuck out, too—as did the crests of other ranges. Forty million years before that, in Eocene time, most parts of the Rocky Mountains seem to have looked much as they do today, and so did the broad basins among them. The region as a whole was closer to sea level, but its relief was essentially the same.

It was a bizarre story, full of odd detail. Limestone, for example, is ordinarily a marine rock, derived from corals and shells. What sort of limestone would form on a surface that came sloping down like a tent roof from the ridgelines of buried mountains? The surface had been laced with streams, Love said, and in the rumpled topography east of the buried ranges the streams filled countless lakes. Old carbonates dissolved and were carried by the streams to the lakes. Lime also leached out of the granite in the mountain cores. Freshwater limestones formed in the lakes, self-certified by the fossils of freshwater snails. There were other fossils as well—discovered in dense compilations in confined areas that have been described as concentration camps and fossil graveyards. They suggest a modern plain in south-central Brazil where heavy seasonal rains so elevate the waters of innumerable lakes that animals crowd up on small islands and perish. Limestone, being soluble in water, forms fertile valleys in eastern North America but in the dry West remains largely undissolved. Its inherent ruggedness holds it high, while weaker rock around it falls away. Limestone is the protective caprock of Pine Bluffs. Junipers were flourishing in it, as were ponderosa pines and Spanish bayonets.

In the Bronco, we moved through the snow toward the mountains, crossing the last of the Great Plains, which had been shaped like ocean swells by eastbound streams. Now and again, a pump jack was visible near the road, sucking up oil from deep Cretaceous sand, bobbing solemnly at its task—a giant grasshopper absorbed in its devotions. As we passed Cheyenne, absolutely all we could see in the whiteout was a raging, wind-whipped flame, two hundred feet in the air, at the top of a refinery tower. "Such a waste," said Love.

Had we been moving west across Wyoming about seventy-five million years ago, in the Campanian age of late Cretaceous time, we would, of course, have been at sea level in the most literal sense. The Laramie Range did not exist, nor did the Bighorns, the Beartooths, the Wind Rivers. There is no evidence of mountains at that

[45]

time anywhere in Wyoming. In an oceangoing boat (which the Bronco in some ways resembled), we would have raised the coastline not far east of Rawlins. Beyond the beach and at least as far as Utah was flat marshy terrain.

Earlier, there had been mountains—a few ranges that were largely in Colorado and poked some miles into Wyoming. They have been called the Ancestral Rockies; but they and the Rockies are scarcely more related than two families who happen at different times to live in the same house. Those Pennsylvanian mountains had worn down flat two hundred and thirty million years before. There had been other mountains as well—in the same region—some hundreds of millions of years before that, in various periods of Precambrian time. The Precambrian evidence, in fact, suggests numerous episodes, across two thousand million years, of the violent rise of big mountains and their subsequent wearing away—any of them as deserving as others to be called ancestral Rockies.

In the middle Precambrian, not long after the end of the Archean eon, lava ran down the sides of big volcanoes and far out onto a seafloor that is now a part of Wyoming. It is impossible to say where the volcanoes stood, but the fact that they existed is stated by the lava. Somewhat later, the lava was folded and faulted, apparently in the making of mountains. Still more Precambrian ranges, of vast dimension, came up in the region, and shed twenty-five thousand feet of sediment into seas that covered parts of Wyoming. After the sediment formed into rock, even more episodes of mountain-building heated and changed the rock: the limestone to marble, the sandstone to quartzite, the shale to slate. Meanwhile, coming into the crust at depths on the order of six miles were vast bodies of fiery-hot magma, much of which happened to have the chemistry of granite. Under eastern Wyoming, where Interstate 80 now crosses the Laramie Range, the magma contained enough iron to tint the feldspar and make the granite pink. It is axiomatic that big crystals grow slowly. Slowly, the magma cooled, forming quartz and feldspar crystals of exceptional size.

All of that occurred in Precambrian time—during the first ninety per cent of the history of the world. Often, Precambrian rock is collectively mentioned as "the basement"—the basement of continents—as if that is all there is to say about it before setting up on top of it the wonders of the world. This scientific metaphor is at best ambiguous—connoting, as it does, in one sense a firm foundation, in another an obscure cellar. In either case, it dismisses four billion years. It attempts to compress the uncompressible. It foreshortens a regional history wherein numerous ancestral mountain ranges developed and were annihilated—where a minor string of Pennsylvanian ridges could scarcely be said to represent the incunabular genealogy of the Rockies.

In late Cretaceous and early Tertiary time, mountains began to rise beneath the wide seas and marsh flats of Wyoming. The seawater drained away to the Gulf of Mexico, to the Arctic Ocean. And, in David Love's summary description, "all hell broke loose." In westernmost Wyoming, detached crustal sheets came planing eastward—rode fifty, sixty, and seventy-five miles over younger rock—and piled up like shingles, one overlapping another. In the four hundred miles east of these overthrust mountains, other mountains began to appear, and in a very different way. They came right up out of the earth. In Love's phrase, they simply "pooched out." Basins flexed between them, filling as they downwarped—folding, too, especially at their edges. These mountains moved, but not much—five miles here, eight miles there. They moved in highly miscellaneous and ultimately perplexing directions. The Wind River Range crept southwest, about five inches every ten years for a million years. The Bighorns split. One part went south, the other east. Similarly, the Beartooths went east and southwest. The Medicine Bows moved east. The Washakies west. The Uintas north. All distances were short, because the mountains were essentially rooted. The Sierra Madre did not move at all. The spines of the ranges trended in as many directions as a weathervane. The Laramie Range trended north-south. The Wind Rivers and Bighorns northwest-southeast. The spectacularly anomalous Owl Creeks and Uintas, lining themselves up at right angles to the axis of the Western cordillera, ran east-west. All these mountain ranges were coming up out of the craton—heartland of the continent, the Stable Interior Craton. It was as if mountains had appeared in Ohio, inboard of the Appalachian thrust sheets, like a family of hogs waking up beneath a large blanket. An authentic enigma on a grand scale, this was one of the oddest occurrences in the tectonic history of the world. It would probe anybody's theories. It happened rapidly. As Dave Love at one point remarked about the Medicine Bow Mountains, "It didn't take very long for those mountains to come up, to be deroofed, and to be thrust eastward. Then the motion stopped. That happened in maybe ten million years, and to a geologist that's really fast."

Twenty thousand feet of rock was deroofed from the rising mountains. The entire stratigraphy from the Cretaceous down to the Precambrian was broken to bits and sent off to Natchez, as the mountains were denuded to their crystalline and metamorphic cores. In half a billion years of history, this was the great event. In the words of the "Geologic Atlas of the Rocky Mountain Region," it was "tectonically unique in the Western Hemisphere and, therefore, it seems to require a somewhat unusual if not unique tectonic interpretation." The foreland ranges, as the mountains east of the overthrust are called (the Wind Rivers, Uintas, Bighorns, Medicine Bows, Laramie Range, and so forth), came into the world with their own odd syncopation, albeit the general chronology went from west to east and the Laramie Range was among the last to rise. "The mountains were restless," Love was saying now. "They didn't all pooch out at once. They moved in fits and starts over a span of time. The Owl Creeks rose in the early Eocene, as did the Uintas. The Medicine Bows, which are farther east, came up before the Uintas. They are all separate mountains with the same general type of origin. They are cohesive in the way that a family is cohesive. They are part of the same event."

The event is known in geology as the Laramide Orogeny. Alternatively it is called the Laramide Revolution.

Mountains always come down, of course, as they

are coming up. In the contest between erosion and orogeny, erosion never loses. For a relatively short time, though, the mountains prevail by rising faster than they are destroyed. In what Love has called "some of the greatest localized vertical displacement known anywhere in the world," the Wind Rivers rose sixty thousand feet with respect to the rock around them, the Uintas fifty thousand, others as much. Frequent rains and many streams helped melt them away. West of Wyoming, in the Eocene, there were no Coast Ranges, no Sierra Nevada. Warm winds off the Pacific brought rains to the Rockies, and a climate similar to the present climate of Florida. In the early Eocene, when the ranges in general looked much as they do today, the mountain-building ceased. In the tectonic quiet, erosion of course continued, and the broad downwarps among the ranges continued to fill.

Then came a footnote to the revolution. "In latest early Eocene, fifty-two million years ago, all hell broke loose again," Love said. From thousands of fissures in northwest Wyoming, lava poured forth by the cubic mile. Torn apart by weather and rearranged by streams, it has since been etched out as the Absaroka Range. "After that, everything went blah," he went on. "In the Oligocene, the tectonic activity was totally dead, and it stayed dead at least until the early Miocene. Thirty million years. Then, in the late Miocene, all hell broke loose again. And all hell has been breaking loose time and again for the last ten million years. This is not a static science."

During those thirty million years after things went blah, the Rockies were quietly buried ever deeper in their own debris—and, not so peacefully, in materials oozing overland or falling from the sky. Much came in on the wind from remote explosive volcanoes—strato-volcanoes of huge size in Idaho, Oregon, Nevada. "And maybe Arizona and California, for all we know," Love said. "Clinical details are still inadequate. By the end of the Eocene, the Washakie and Owl Creek Mountains were so deeply buried that the Wind River and Bighorn Basins had coalesced above them. At the end of the Oligocene, only a thousand to four thousand feet of the highest mountains protruded above the aggradational plain. Streams were slow and sluggish and so choked with ash they were unable to erode."

Rhinoceroses lived through those changes, and ancestral deer and antelope, and horses that had three toes and were by now the size of collies. As altitude and aridity increased, a subtropical world of figs, magnolias, and breadfruit cooled into forests of maple, oak, and beech. Altitude alone could not account for the increasing coolness. It foreshadowed the coming ice.

The burial of the mountains continued far into the Miocene, with—as Love described it—"surprising thicknesses of sandstone and tuffaceous debris." Volcanic sands, from Yellowstone and from elsewhere to the west, were spread by the wind, and in places formed giant dunes. Two thousand feet of sand accumulated in central Wyoming. Nineteen thousand—the thickest Miocene deposit in America—went into the sinking Jackson Hole. From the Wind River Mountains southward to Colorado and eastward to Nebraska, the plain was unbroken except for the tops of the highest peaks. Rivers were several thousand feet higher than they are now. The ranges, buried almost to their summits, were separated by hundreds of miles of essentially flat terrain. Mountains that were completely covered—lost to view somewhere below the water-laid sediments and deep volcanic sand—outnumbered the mountains that barely showed through. At its maximum, the broad planar surface occupied nearly all of Wyoming—upward of ninety per cent—and on it meandered slow streams, making huge bends and oxbows. As events were about to prove, the deposition would rise no higher. This—in the late Miocene—was the level of maximum fill.

For something began to elevate the region—the whole terrain, the complete interred family of under-thrust, upthrust, overthrust mountains—to lift them swiftly about a mile. "The uplift was not absolutely uniform everywhere," Love said. "But nothing ever is." What produced this so-called epeirogeny is a subject of vigorous and sometimes virulent argument, but the result, continuing to this day, is as indisputable as it has been dramatic. It is known in geology as the Exhumation of the Rockies.

From around and over the Wyoming ranges alone, about fifty thousand cubic miles have been dug out and taken away, not to mention comparable excavations in the neighboring cordillera. Though the process has been going on for ten million years, it is believed to have been particularly energetic in the past million and a half, in part because of the amount of rain that fell on the peripheries of continental ice. In response to the uplift, the easygoing streams that had aimlessly wandered the Miocene plain began to straighten, rush, and cut, moving their boulders and gravels in the way that

chain saws move their teeth. The streams lay in patterns
that had no relationship to the Eocene topography
buried far below. Some of them, rushing along through
what is now the Wyoming sky, happened to cross the
crests of buried ranges. After they worked their way
down to the ranges, they sawed through them. Some
effects were even odder than that. If a river happened
to be lying above a spur of a buried range, it would cut
down through the spur, and seem, eventually—without
logic, with considerable magic—to flow into a mountain range, change its mind, and come back out another
way. "Eventually," of course, is now. The North Platte
River now flows into the rocks of the Medicine Bow
Mountains, comes out again to cross the Hanna Basin,
and then runs through the Seminoe Mountains and the
Granite Mountains. It is joined by the Sweetwater River
on the crest of the Granite Mountains. Irrespective of
modern topography, the pattern of the rivers is Miocene. On the Laramie Plains, the Laramie River behaves
for a while in a deceptively conventional manner. It
establishes itself as the centerpiece of the basin, pre-
tending to be the original architect of the circumvallate
scene, but then takes a sharp right and, like a bull with
lowered head, charges the Laramie Range. The canyon
it has made is deep and wild. Water roars through it.
When, in the exhumation, the river got down to the
mountains, it packed the abrasive power to cut them
in half.

In fact, there is no obvious relationship between
most of the major rivers in Wyoming and the landscapes
they traverse. While rivers elsewhere, running in their
dendritic patterns like the veins in a leaf, shape in

harmony the landscapes they dominate, almost all the rivers of the Rockies seem to argue with nature as well as with common sense. At Devil's Gate, on the Oregon Trail, the Sweetwater River flows into a hill of granite and out the other side. The Wind River addresses itself to the Owl Creek Mountains and flows right at them. It, too, breaks through and comes out the other side. It, too, flowed across the totally buried mountains in the Miocene, and descended upon them during the exhumation. The anomaly is so startling that early explorers, and even aborigines, did not put one and one together. To the waters on the south side of the mountains they gave the name Wind River. The waters on the north side they called the Bighorn. Eventually, they discovered Wind River Canyon.

On the east flank of the Laramie Range is a piece of ground that somehow escaped exhumation. Actually contiguous with Miocene remains that extend far into Nebraska, it is the only place between Mexico and Canada where the surface that covered the mountains still reaches up to a summit. To the north and south of it, excavation has been deep and wide, and the mountain front is of formidable demeanor. Yet this one piece of the Great Plains—extremely narrow but still intact —extends like a finger and, as ever, touches the mountain core: the pink deroofed Precambrian granite, the top of the range. At this place, as nowhere else, you can step off the Great Plains directly onto a Rocky Mountain summit. It is known to geologists as the gangplank.

Now the Bronco began to rise through the snow, and Love remarked that we were on the gangplank. The land fell away on either side, and in the low visibility we seemed indeed to be on a plank going up into the sky. As we continued to climb, the strip of earth became narrower and narrower. We pulled over onto the shoulder, shut off the motor, and squinted. We appeared to be on a bridge—built of disassembled Rockies and travelled ash—crossing a great excavation through flapping veils of snow. "There are twelve inches of precipitation per annum here, and it's mostly snow," he said. "The mean temperature is thirty-eight degrees. The growing season is less than ninety days. Conditions are about the same in this part of Wyoming as at the Arctic Circle." With that, we gave up the geology and crawled off to his home in Laramie, defeated by the snow. We went back to the gangplank in clear weather.

It was half a mile long. To the north and south, the land fell away along the mountain front in profound excavation of the sediments that once had been there. The excavation had exposed the broken, upturned ends of Pennsylvanian sandstones, dipping steeply eastward and leaning on the mountains. They rested there like lumber stood against a barn. These red sandstones lean against the Laramie Range on both sides. By themselves, they tell the story of the Laramide Orogeny, for they are a part of what was deroofed. They are a part of the Paleozoic package that once rested flat on the deep Precambrian granite. They are thought by some to have been Pennsylvanian beach sands. Whatever they may have been, they were indubitably horizontal, and for roughly two hundred and fifty million years remained horizontal while layer after layer of sediment accumulated above them, finally including the floors of the Cretaceous seas. Then all hell broke loose, and the granite rose beneath them. The granite core came up like a basement elevator that rises through a city sidewalk, pushing to either side a pair of hinged doors. That was the chronology of numerous ranges—the old hard stuff from far below breaking upward through roofrock and ultimately standing highest, while the ends of the roofrock lean on the flanks in gradations of age that are younger with distance from the core. The broken ends of that Pennsylvanian sand—the outcropping edges of the tilted strata—had weathered out as a rough, serrate ridge along the border of the range. Rocky Mountain ranges are typically flanked by such hogbacks. Boulder, Colorado, is backdropped with hogbacks (the Flatirons), which are more of the same Pennsylvanian strata leaning against the Front Range. Now, on the gangplank, Love said parenthetically, "You are seeing Paleozoic rocks for the first time since the Mississippi River. They go all the way through—under Iowa and Nebraska—but they're buried."

In the fall of 1865, Major General Grenville Dodge and his pack trains and cavalry and other troops were coming south along the St. Vrain Trail, under the front of the Laramie Range. The Powder River campaign, behind them, had been, if not a military defeat, a signal failure in its purpose: to cow the North Cheyennes and the Ogallala Sioux. General Dodge, though, was preoccupied with something else. President Lincoln, not long before he died, had instructed Dodge to choose a route for the Union Pacific Railroad. Dodge, like others before him, had sought the counsel of Jim Bridger, the much celebrated trapper, explorer, fur trader, commercial entrepreneur, and all-around mountain man.

Bridger, who was sixty by then, had preceded almost
everybody else into the West by two or three decades
and knew the country as few other whites ever would.
It was he who discovered the Great Salt Lake, report-
ing his find as the Pacific Ocean. It was he whose
descriptions of Jackson Hole, Yellowstone Lake, Yellowstone Falls, the Fire Hole geysers, and the Madison
River had once been known as "Jim Bridger's lies." His
father-in-law was Chief Washakie. And now this blue-
coat general wanted to know where to put a railroad.
The Oregon Trail went around the north end of the
Laramie Range and up the Sweetwater to South Pass—
to say the least, an easy grade. But for a competitive
transcontinental railroad the Sweetwater was a route
of wide digression and no coal. Bridger mentioned
Lodgepole Creek and said the high ground above it was
the low point on the crest of the Laramie Range (a fact
that theodolites would in time confirm). The route could
go there.

So Dodge, in 1865, coming south from the Powder
River, left his pack trains and cavalry on the St. Vrain
Trail and led a small patrol up Lodgepole Creek. At the
top, he turned south and did reconnaissance of the summit terrain. In the small valley of a high tributary of
Crow Creek—five or ten miles south of Bridger's recom-
mendation—he surprised a band of Indians. His report
does not say of what tribe. They were hostiles—or at
least became so after Dodge started firing at them. At
the moment of mutual surprise, they were between him
and his main column, and that made him tactically
nervous. The patrol dismounted and walked due east—
"holding the Indians at bay, when they came too near,

with our Winchesters." In this manner, the gangplank was discovered. As Dodge kept going east, expecting to reach the escarpment from which he would signal with smoke, he reached no escarpment. Instead, he reached the remnant of the high ancient surface—this interfluvial isthmus between Crow Creek and Lone Tree Creek—touching the mountain summit.

It led down to the plains without a break. I then said to my guide that if we saved our scalps I believed we had found the crossing.

General Dodge went back east, and in the spring of 1867 returned with his route approved. The Union Pacific at that time ended in the middle of Nebraska. He got off the train, went up the North Platte, up the Lodgepole, and, as he approached the mountains, went directly overland to Crow Creek, where he staked out the western end of the railroad's next division. Without much pleasing anybody, he named the place Cheyenne. In no time, he was defending himself against furious Cheyennes. They killed soldiers and laborers, pulled up survey stakes, stole animals, and destroyed equipment. When some politicians, bureaucrats, and financiers arrived on a see-it-yourself junket west, the Cheyennes attacked them. With drawn revolver, General Dodge told his visitors, "We've got to clean these damn Indians out or give up building the Union Pacific Railroad. The government may take its choice."

The narrowest point on the gangplank is wide enough for the Union Pacific and nothing else. The interstate highway clings to one side. The tracks and lanes

are so close that the gangplank resembles the neck of a
guitar. A long coal freight slid by us. "The coal isn't
piled higher than the tops of the gondolas," Love commented. "It's an environmental move—to keep the dust
from blowing downwind." He said it was a good idea,
no doubt, but he had experienced so many cinder
showers earlier in his life that he could not help thinking that this latter-day assault on dust was "like bringing a fire under control at timberline." A cinder shower
was what happened when an old-time locomotive pulled
into a town and blew its stack. He also said that this
could not have been an important emigrant route, because there was a lack of grass and water—absolute
necessities for animal-powered travel. To the Union
Pacific, however, the gangplank offered speed, effi-
ciency, and hence predominance with respect to the
competition. When the Denver & Rio Grande was labor-
ing up switchbacks in a hampering expenditure of money
and time—and the Santa Fe was struggling not only
with mountains but also with desert terrain—the Union
Pacific had already run up the gangplank, opened the
West, and become everybody's Uncle Pete. Love said,
"Out here, Uncle Sam is a gnat under a blanket compared to Uncle Pete. The Union Pacific had the best of
it. This Miocene Ogallala formation was the youngest
of the high-plains deposits that lapped onto the mountain front. It's subtle and seems academic until you try
to build a railroad. This is the only place in the whole
Rocky Mountain front where you can go from the Great
Plains to the summit of the mountains without snaking
your way up a mountain face or going through a tunnel.
This one feature had more to do with the building of

the West than any other factor. I don't diminish the importance of the Oregon Trail, but here you had everything going for you. This point hasn't been made before."

When the railroad was built, it was given (by the federal government) fifty per cent of the land in a forty-mile swath along its route—in checkerboard fashion, one square mile in every two. Today, Uncle Pete owns, among many things, the Rocky Mountain Energy Company, the Upland Industries Corporation, the Champlin Petroleum Company, and enough unmined uranium to send Wyoming to the moon. In Cheyenne, the Union Pacific station and the state capitol face each other at opposite ends of Capitol Avenue. The Union Pacific station came out of the Laramie Range, forty miles west, and, like the range itself, is sheathed in the russet Pennsylvanian sandstone and has a foundation of Precambrian granite. At least as imposing as the capitol, it is a baronially escutcheoned mountain of grandeur.

Indians, of course, had used the gangplank for who knows how long before General Dodge surprised them on the Laramie summit. They had crossed it on their journeys from the Great Plains to the Laramie Basin and on up to hunting grounds in the Medicine Bow Mountains. And the Indians, from the beginning, were themselves following a trail. Buffalo discovered the gangplank. "It was a buffalo trail," Love said. "Buffalo were the real trailmakers—trails you wouldn't believe. They were as good as the best civil engineers. It remains true today. If you're in Yellowstone, in the backcountry, and you have trouble finding your way across swamps, mountains, and thermal areas, you look for a buffalo trail and you'll get through." Beside Interstate 80 on the gangplank, a sign said, "GAME CROSSING."

༄ We moved off the gangplank and into a highway throughcut of pink granite. Love said, "Now we are on the mountain, on the Precambrian core. You have to watch closely. This fantastic geology is subtle. I-80 was not built to show it off but to take advantage of its beneficences." There were more pink granite cuts and also some dark, shattered amphibolite that had been the country rock into which the granite intruded 1.4 billion years ago. The interstate had sliced through a section where the bright-pink granite and the charcoal-gray amphibolite met. It was as if a wall painter had changed colors there. The dark rock was full of fracture planes and cleavage planes. "That rock probably had been messed around for a long time before the granite came," Love said. "It could be two, three billion years old. We don't know."

As our altitude increased, the granite roadcuts became deeper and higher and seemingly more rutilant. The rock was competent. There were no benches, and the cuts were as much as fifty metres high. Resembling marbled steak, they were shot through with veins of quartz, where, long after the granite formed, it cracked and quartz filled it in. The walls were indented with vertical parallel grooves, like giant wormtrails in some exotic sediment. These were actually fossil shot holes and unloaded guide holes from the process of pre-splitting. The highway builders drilled the holes and then dynamited one of three. In this manner, they—and we—reached eight thousand six hundred and forty feet, the highest point on Interstate 80 between the Atlantic and the Pacific. What appeared to be the head of a chicken sat at the top of a big granite block, as if it had been chopped off there. Only when we drew close did I glance up and see that it was Abraham Lincoln. It was, in fact, an artful likeness, resting on an outsized plinth. Years ago, this had been the summit of the Lincoln Highway, which was now incorporated in its substance, if not in its novel spirit, into the innards of the interstate.

We left I-80 there and bucked the southwest wind, crossing the surprisingly flat mountain-crest terrain on a pair of ruts in the pink granite, which had crystals the size of silver dollars. The view from that high wide surface took in a large piece of the front of the Rockies, with the Never Summer Mountains standing out clearly in Colorado, to the south, and, to the west, the bright peaks of the Snowy Range. The Snowy Range—rising white above a dark high forest—appeared to be on top of the Medicine Bow Mountains. Remarkable as it seemed, that was the case. At the ten-thousand-foot level, the bottom of the Snowy Range rests on the broad flat top of the Medicine Bows like a sloop on water, its sails flying upward another two thousand feet. In the Miocene, the high flat Medicine Bow surface at the base of the Snowy Range was the level of maximum fill. In the fifty miles between the Snowy Range and our position on top of the Laramie Range lay the gulf of the excavated Laramie Plains. Our line of sight to the tree line of the Medicine Bows had been landscape in the Miocene. From twelve thousand feet it had gently sloped to about nine thousand where we were, and as we turned and faced east and gazed on down that mostly vanished plane we could all but see the Miocene surface continuing—as Love expressed it—"on out to East G-string."

Everywhere in the central Rockies, that highest level of basin fill touched the eminent ranges at altitudes that are now between ten and twelve thousand feet with results that are as beautiful as they are anomalous in the morphology of the world's mountains. In the Beartooths, for example, you can ascend a glacial valley that—in its U shape and high cirques—closely resembles any hanging valley in the Pennine Alps; but after you climb from ten to eleven to twelve thousand feet you do not find a Weisshorn fingering the sky. Instead, you move into an unexpectable physiographic setting, which, after steep slopes above a dry Wyoming basin, is lush and paradisal to the point of detachment from the world. Alpine meadows with meandering brooks are spread across a rolling but essentially horizontal scene, in part forested, in part punctuated with discrete stands of conifers and small cool lakes. The Medicine Bows are also like that—and the Uintas, the Bighorns. Their high flat surfaces, with peaks that seem to rest on them like crowns on tables, make no sense unless—as you look a hundred miles from one such surface to another across a deep dividing basin—you imagine earth instead of air: the Miocene fill, the continuous terrain. The high plateaus on the shoulders of the ranges, remaining from that broad erosional plane, have been given various names in the science, of which the most prominent at the moment is subsummit surface. "There's a plateau above Union Pass in the Wind River Range that's twelve thousand feet and flatter than a turd on a hot day," Love recalled, and went on to say that at such an altitude in flat country he sometimes becomes panicky—which does not happen if he is among craggy peaks, and seems to be a form of acrophobia directly related to the oddity of being in southern Iowa at twelve thousand feet.

With those big crystals, the granite under our feet was about as coarse as granite ever gets, and, as a result, was particularly vulnerable to weather. Its pink feldspar, black mica, and clear glass quartz had been so exposed there for millions of years that gravels could be scraped off without the help of dynamite. The Union Pacific took advantage of this, ballasting its roadbed with pink granite for eight hundred miles. There was almost no soil in that part of the range—just twelve miles' breadth of rough pink rock. "As you go from Chicago west, soil diminishes in thickness and fertility, and when you get to the gangplank and up here on top of the Laramie Range there is virtually none," Love said. "It's had ten million years to develop, and there's none. Why? Wind—that's why. The wind blows away everything smaller than gravel."

Standing in that wind was like standing in river rapids. It was a wind embellished with gusts, but, over all, it was primordially steady: a consistent southwest wind, which had been blowing that way not just through human history but in every age since the creation of the mountains—a record written clearly in wind-scored rock. Trees were widely scattered up there and, where they existed, appeared to be rooted in the rock itself. Their crowns looked like umbrellas that had been turned inside out and were streaming off the trunks downwind. "Wind erosion has tremendous significance in this part of the Rocky Mountain region," Love said. "Even down in Laramie, the trees are tilted. Old-timers used to say that a Wyoming wind gauge was an anvil on a length of chain. When the land was surveyed, the surveyors couldn't keep their tripods steady. They had to work by night or near sunrise. People went insane because of the wind." His mother, in her 1905 journal, said that Old Hanley, passing by the Twin Creek school, would disrupt lessons by making some excuse to step inside and light his pipe. She also described a man who was evidently losing to the wind his struggle to build a cabin:

He was putting up a ridgepole when the wind was blowing. He looked up and saw the chipmunks blowing over his head. By and by, along came some sheep, dead. At last one was flying over who was not quite gone. He turned around and said, "Baa"—and then he was in Montana.

Erosion, giving the landscape its appearance, is said to be the work of water, ice, and wind; but wind is, almost everywhere, a minimal or negligible factor, with exceptional exceptions like Wyoming. Looking back

[65]

across the interstate—north up the crest of the range—
among ponderosas, aspens, and limber pines we could
see the granites of Vedauwoo Glen, which had weathered out in large blocks, as granite does, along inter-
secting planes of weakness, while windborne grit had
rounded off the corners of the blocks. Where some had
tumbled and become freestanding, grit flying close
above the ground had abraded them so rigorously that
the subsummit surface was, in that place, a flat of giant
mushrooms. The cliffs behind them also looked organic
—high piles of rounded blocks, topped in many places
by narrowly balanced boulders that were undercut almost to the point of falling. Love, contemplative, appeared to be puzzling out some deep question in geo-
morphology. At length, he said, "When wild horses
defecate, they back up to a place where other wild
horses have defecated, and so on, until they build turd
towers, like those, in the air. Domestic horses do not
do this."

At the Wyoming Information Center, beside Interstate 80 just south of Cheyenne, eleven picnic tables are
enclosed in brick silos, and each silo has a picture
window, so that visitors to Wyoming can picnic more
or less al fresco and not be blown home. On the range,
virtually every house has a shelter belt of trees—and
for the most part the houses are of one story. Used
tires cover the tops of mobile homes. Otherwise, wind
tears off the roofs. Mary Kraus, a sedimentologist from
the University of Colorado, got out of her car one day
in north-central Wyoming and went to work on an outcrop. The wind blew the car off a cliff. A propeller-
drawn airplane that serves Wyoming is known as the

Vomit Comet. When people step off it, they look like spotted slate.

"Most people today don't realize the power of wind and sand," Love said. "Roads are paved. But in the first fifty years of the Lincoln Highway you didn't like to travel west in the afternoon. You'd lose the finish on your car. Your windshield became so pitted you could hardly see out." The Highway Department has not yet paved the wind. On I-80, wind will capsize tractor-trailers. When snow falls on Wyoming, its travels are only beginning. Snow snows again, from the ground up, moves along the surface in ground blizzards that can blind whole counties. Ground blizzards bury houses. In roadcuts, they make drifts fifty feet deep. The wind may return ahead of the plows and take the snow away. The old-timers used to say, "Snow doesn't melt here; it just wears out." Interstate 80 has been closed by snow in Wyoming in every month but July and August—sometimes closed for days. It is known as the Snow Chi Minh Trail. Before Amtrak dropped its Wyoming passenger service, people stranded on the Snow Chi Minh Trail used to abandon their cars and make their escape by train. The most inclement stretch of 80 is east of Rawlins where it skirts the tip of the Medicine Bows, where anemometers set on guardrails beside the highway frequently catch the wind exceeding the speed limit.

Now, looking from mountains to mountains west over the Laramie Plains—his gaze bridging fifty miles of what had fairly recently been solid ground—Love said he thought the role of the wind had been much greater than hitherto suspected in the Exhumation of the Rockies. Water, of course, was the obvious agent for the digging and removal of the basin fill, as a look at the Mississippi Delta would tend to confirm. Many miles off the coast there, you could drill down into the muck and after fifteen thousand feet the bit would still be in the Miocene. He continued, "We know, however, the approximate volume of sediment from the Powder River Basin, the Bighorn Basin, the Wind River Basin, the Laramie Basin, and so forth. We can say it all went downhill to the Mississippi Delta. But go to the delta. Look at the volumes. There's an enormous discrepancy. You add up what's down there in the Gulf and what was removed here, and they don't square. A great deal more has been removed from here than is down there. Streams only account for about half the material that was taken up and out of here. Since it is not all in the delta, where did it go? So much has been taken away that it's got to be explained in some other manner. I think the wind took it. My personal feeling is that a lot of it blew eastward to the Atlantic. Possibly some went to Hudson Bay. We don't know. These are problems we are trying to grapple with at the present time. How much did the wind take? Again, we don't know, but in one dust storm several years ago a great deal of debris from Kansas and Nebraska and Colorado went into the Atlantic—a storm that lasted only a couple of days."

Such storms are frequent, and this one was not unusual in size or duration. It is noteworthy because its effects were studied and published, in the *Journal of Sedimentary Petrology*. When the dust appeared above the coast of Georgia—as thick haze—it attracted the attention of researchers at the Skidaway Institute of

Oceanography, near Savannah. The cloud of particles was two miles in height, and satellite photographs showed its other dimensions: four hundred thousand square miles. With air-sampling-and-measuring equipment, the Skidaway people collected particles. They reported in that one storm enough dust to account for twenty-five per cent of the annual rate of sedimentation —from all rivers as well as the air—in the proximate North Atlantic. Moreover, about eighty-five per cent of it was a clay mineral called illite. Silts coming out of East Coast rivers include very little illite, and yet illite is predominant among the sediments of the ocean floor. By Skidaway's calculations, that one storm's deposits in the ocean amounted to a million tons.

Moving even farther from the interstate on the sub-summit surface, we came upon a granite pyramid, sixty feet high, sixty feet wide at the base. It had been designed by the architect H. H. Richardson and weighed six thousand tons—enough to prevent its blowing over. We stood in its lee. The wind was coming in pulses that made percussions in the ears. The incongruity of this monument was in direct proportion to its stark isolation. It was Uncle Pete's version of Interstate 80's Abraham Lincoln. It commemorated the brothers Oakes and Oliver Ames—Massachusetts shovel-makers, railroad financiers—whose Crédit Mobilier of America made construction contracts with itself in enjoying the fruits of subsidy of the Union Pacific Railroad. If you belonged to the United States Congress, you could buy shares of Crédit Mobilier stock for fifty per cent of their value. Near the apex of the east side of the pyramid was Oliver's face in a portrait plaque, sculptured in 1881 by Augustus Saint-Gaudens, whose William Tecumseh Sherman stands in Manhattan's Grand Army Plaza and Robert Gould Shaw in Boston Common. Saint-Gaudens' plaque of Oakes Ames was on the west side of the pyramid, facing the wind. The monument had been built beside the Union Pacific at the railroad's highest point, but the railroad's highest point was somewhere else now; the alignment had been changed in 1901, and the track was three or four miles away. The original roadbed had become so indistinct that a geologist was required to point out where it had been, which he did. Oakes' nose had been shot off with a high-powered rifle. Oliver's nose had been shot away, too, and a large part of his face. Love remarked that Greek, Roman, and Saracen vandals broke off the noses from pieces of sculpture. Probably the Vandals did, too.

Back on the interstate and just west of Abraham Lincoln, the rock became younger again, as we left the Precambrian range core and encountered the same Pennsylvanian red sandstone that had leaned on the mountain on the other side. It was rich red, and the cuts were very big as the road plunged through them in christie turns, running down the mountains through Telephone Canyon. Somewhere overhead had been the first telephone wire ever strung across the Rockies. The President of the United States, with a dozen horses and companions, rode up Telephone Canyon on his way to Cheyenne in 1903. His mustache was an airfoil with a fineness ratio that must have impressed the Wright brothers. He wore a three-gallon hat. His paunch at the time was under control. The interstate trail was more than a little wild then, but manifestly so was he. The red rock is of so much beauty there, and competence, that people collect it for building material, banging it free from the shattered roadcuts and loading it into pickups, much as ranchers did when they first came to the Laramie Plains and ascended the mountains in wagons and collected the rock to build their homes. It is a porous and permeable, fine-grained, hard, brittle sandstone; and because it rests on impermeable granite water moves through it downhill. Released in a fault zone at the bottom, the water leaps to the surface in artesian fountains—the springs that established Laramie. The bright-red roadcuts, ten and twenty metres high, were capped with a buff-colored limestone, which had been deposited in tropical waters on top of the Pennsylvanian sand. After a mountain range rises under layers of flat-lying rock and bends them upward until they all but stand on end, the slopes of the eroding mountains will descend more gently than the dip of the molested strata. And so, as we plunged down Telephone Canyon, the interstate was tilting less than the rock of the roadcuts, and the red sandstone yielded gradually, interstitially, to the younger limestones, until the sandstone was gone altogether and we were moving through the floor of an ocean. It was full of crinoids, brachiopods, and algal buttons, which had lived near the equator in a place like the Bismarck Archipelago or an arm of the Celebes Sea.

The canyon opened to the plains—a broad dry sea of the interior Rockies—and soon we were on Grand Avenue, Laramie, passing the University of Wyoming, whose buff buildings on wide soft lawns could never be said to resemble roadcuts, notwithstanding the crinoids in their walls, the brachiopods and algal buttons. We passed Love's home, on Eleventh Street, and his U.S.G.S. office—a building on the campus, adjacent to the Geology Department and to a life-size two-story sculpture of Tyrannosaurus rex, the toughest-looking critter in the history of the earth, a native, needless to say, of Wyoming. We passed St. Matthew's Episcopal Cathedral, which also—as Love had reason to regret—contained in its walls brachiopods, crinoids, and algal buttons. He once taught Sunday school there. He took the kids outside and showed them the fossils in the church walls. He described the environment in which the creatures had lived. He mentioned the age of the rock. He explained how things evolve and the fit prosper. Here endeth his career in sedimentary theology.

A few miles north of town, we passed the quarry out of which had emerged not only the university and St. Matthew's Cathedral but also the Ivinson Home for Aged Ladies and the Albany County Courthouse. "It's a limey sandstone, slightly fossiliferous," he said. "It holds up pretty well." We continued north along the foot of the Laramie Range and then turned east into the mountains, climbing a canyon downsection until we had returned to Precambrian time. The rock in this place was even older than the neighboring subsummit granite, and some of it was chatoyant: flashing like a cat's eye. It flashed every color in the spectrum. The rock was anorthosite, nearly fifteen per cent aluminum, Love said. When the bauxites of the Caribbean run out, anorthosite will be a source of aluminum. "Anorthosite is tough, has a high melting point, and doesn't fracture easily," he continued. "Hence it might be useful for containing atomic waste." Anorthosite is rare on earth. It began forming during the Archean eon and predominantly dates from an age of the later Precambrian known as Helikian time. Yet the high Adirondacks are largely anorthosite. The choice they present is to seal up our spent nuclear fuel inside them or dismantle them one at a time to make beer cans. Anorthosite is more plentiful elsewhere. It is most of what you are looking at when you are looking at the moon.

Moving on west, another day, we crossed the Laramie Plains on I-80 through a world of what to me were surprising lakes. They were not glacial lakes or manmade lakes or—as in Florida—sinkhole lakes filling bowls of dissolved limestone. For the most part, they had no outlets, and were therefore bitter lakes—some alkaline, some saline, some altogether dry. Of Knadler Lake, about a mile long, Love said, "That's bitter water —sodium sulphate. It would physic you something awful." A herd of twenty antelopes galloped up the shore of Knadler Lake. Most of the lakes of the world are the resting places of rivers, where rivers seek their way through landscapes that have been roughed up and otherwise left chaotic by moving ice. Ice had never covered the Laramie Plains. What, then, had dug out these lakes?

Love's response to that question was "What do you suppose?"

We had seen—a mile or two away—a hole in the ground eleven miles long, four miles wide, and deeper than the Yellow Sea. There were some puddles in it, but it did not happen to intersect any kind of aquifer, and basically it was dry. With a talent for understatement, the people of Laramie call it the Big Hollow. Geologists call it a deflation basin, a wind-scoured basin, or—more succinctly—a blowout. The wind at the Big Hollow, after finding its way into some weak Cretaceous shales, had in short order dug out four million acre-feet and blown it all away. Wind not only makes such basins but maintains them—usually within frameworks of resistant rock. On the Laramie Plains, the resistant rock is heavy quartzite gravel—Precambrian pieces of the Snowy Range which were brought to the plains as the beds of Pleistocene rivers. Wet or dry, all the lakes we passed had been excavated by the wind. It was a bright cloudless morning with a spring breeze. Spheres of tumbleweed, tumbling east, came at us on the interstate at high speed, like gymfuls of bouncing basketballs dribbled by the dexterous wind. "It's a Russian thistle," Love said. "It's one of nature's marvels. As it tumbles, seeds are exploded out."

Across the green plains, the Medicine Bow Mountains and the Snowy Range stood high, sharp, and clear, each so unlike the other that they gave the impression of actually being two ranges: in the middle distance, the flat-crested Medicine Bows, dark with balsam, spruce, and pine; and, in the far high background, the white and treeless Snowy Range. That the one was in fact directly on top of the other was a nomenclatural Tower of Babel that contained in its central paradox the narrative of the Rockies: the burial of the ranges, the subsequent uplifting of the entire region, the exhumation of the mountains. As if to emphasize all that, people had not only named this single mountain range as if it were two but also bestowed upon the highest summit of the Snowy Range the name Medicine Bow Peak. It was up there making its point, at twelve thousand thirteen feet.

We passed a stone ranch house a century old, and a set of faded ruts in the rangeland that were older than the house. This was the Overland Trail, abandoned in 1868 after seven dismal years. "A nasty route," Love remarked. "Steep grades. Many rocks. Poor water. Poor grass. It was three days across the Laramie Plains at ten miles a day. It was often muddy and boggy. A disaster."

When, in the orogeny, the Medicine Bow Mountains were shoved a few miles east, the rock in front of them folded, in the way that a tablecloth will fold if you push at it with your hands. The anticlines among the folds formed traps for migrating fluids. All about us were pump jacks bobbing for oil.

Boulder beds in the roadcuts represented, as Love put it, "the deroofing of the Medicine Bow Mountains in the first pulsation of the Laramide Revolution." The beds were of Paleocene age. In a knife-edge ridge a few miles farther on, the interstate had exposed the same conglomerates tilted forty-five degrees as mountain-building continued. And soon after that came a flat-lying Eocene deposit. "So you have a time frame for the orogeny," Love said, and this was when he added, "It didn't take very long for those mountains to come up, to be deroofed, and to be thrust eastward. Then the motion stopped. That happened in maybe ten million years, and to a geologist that's really fast."

Near Arlington, an anomalous piece of landscape reached straight out from the mountains like a cause-way heading north. It was capped with stream gravel, brought off the mountains by furious rivers rushing through the tundras of Pleistocene time. The gravel had resisted subsequent erosion, while lighter stuff was washed away on either side. Geologists call such things pediments, and Love remarked that the one before us was "the most striking pediment in this region." In my mind's eye I could see the big braided rivers coming off the Alaska Range, thickly spreading gravel a mile wide, perhaps to preserve beneath them the scenes of former worlds. Where I-80 cut through the Arlington pediment, the Pleistocene gravel rested on Eocene sandstones, on red and green claystones; and they in turn covered conglomerates that came from the mountains when the mountains were new. One could read upward from one world to another: the boulders falling from rising mountains, the quiet landscapes after the violence stopped— all preserved in a perplexing memento from the climate of an age of ice.

In a cut eight miles farther on, that early conglomerate was in contact with Cretaceous rock bent upward even more steeply as the Laramide Orogeny lifted the mountains. Picking through the evidence in the conglomerate was like sorting out debris from an explosion. One after another, I chose a cobble from the roadcut, handed it to Love, and asked him what it was. A Paleozoic quartzitic sandstone, for example—probably Mississippian. Grains rounded. No biotite. In fact, no mica of any kind. A Cretaceous sandstone. That would be from nearby, not from the mountains. A Paleozoic or Precambrian chert. Some Hanna formation sandstone, Paleocene in age—the matrix of the conglomerate. Some Precambrian quartzite from the Snowy

Range, two billion years old. Some bull quartz from a vein in the Precambrian. And one he didn't know.

While the orogeny was making mountains, it was also making basins, for which it is less noted, even where the basins are a good deal deeper than Mt. Everest is high. As we crossed the Medicine Bow River and approached the North Platte and Rawlins, we moved out upon the surface of the Hanna Basin. It was choppy but essentially level nondescript ground, like all the rest of the rangeland on the apron of the mountains. It was not water, and we were not in a boat, but in some ways it seemed so as we crossed a basin forty-two thousand feet deep. It is the deepest structural basin in North America. It is Cretaceous, Paleocene, and Eocene rock, bent in U's, with seams of coal as much as fifty feet thick in the arms of the U's. Union Pacific.

We crossed the North Platte, climbed some long grades, examined a few roadcuts, and pulled off on the shoulder at Rawlins to absorb, in the multiple exposures of the Rawlins Uplift, its comprehensive spread of time —Rawlins, where his mother had boarded the stage north, three-quarters of a century before.

In the United States Geological Survey's seven-and-a-half-minute series of topographic maps is a quadrangle named Love Ranch. The landscape it depicts lies just under the forty-third parallel and west of the hundred-

[77]

and-seventh meridian—coordinates that place it twelve miles from the geographic center of Wyoming. The names of its natural features are names that more or less materialized around the kitchen table when David Love was young: Corral Draw, Castle Gardens, Buffalo Wallows, Jumping-Off Draw. To the fact that he grew up there his vernacular, his outlook, his pragmatic skills, and his professional absorptions about equally attest. The term "store-bought" once brightened his eyes. When one or another of the cowpunchers used a revolver, the man did not so much fire a shot as "slam a bullet." If a ranch hand was tough enough, he would "ride anything with hair on it." Coffee had been brewed properly if it would "float a horseshoe." Blankets were "sougans." A tarpaulin was a "henskin." To be off in the distant ranges was to be "gouging around the mountains." In Love's stories of the ranch, horses come and go by the "cavvy." If they are unowned and untamed, they are a "wild bunch"—led to capture by a rider "riding point." In the flavor of his speech the word "ornery" endures.

He describes his father as a "rough, kindly, strong-willed man" who would put a small son on each knee and—reciting "Ride a cockhorse to Banbury Cross to see a fine lady upon a white horse"—give the children bronco rides after dinner, explaining that his purpose was "to settle their stomachs." Their mother's complaints went straight up the stovepipe and away with the wind. When their father was not reciting such Sassenach doggerel, he could draw Scottish poems out of the air like bolts of silk. He had the right voice, the Midlothian timbre. He knew every syllable of "The Lady of the

Lake." Putting his arms around the shoulders of his wee lads, he would roll it to them by the canto, and when they tired of Scott there were in his memory more than enough ballads to sketch the whole of Scotland, from the Caithness headlands to the Lammermuir Hills.

David was fifteen months younger than his brother, Allan. Their sister, Phoebe, was born so many years later that she does not figure in most of these scenes. They were the only children in a thousand square miles, where children outnumbered the indigenous trees. From the ranch buildings, by Muskrat Creek, the Wind River Basin reached out in buffalo grass, grama grass, and edible salt sage across the cambered erosional swells of the vast dry range. When the wind dropped, this whole wide world was silent, and they could hear from a great distance the squeak of a horned lark. The nearest neighbor was thirteen miles away. On the clearest night, they saw no light but their own.

Old buffalo trails followed the creek and branched from the creek: old but not ancient—there were buffalo skulls beside them, and some were attached to hide. The boys used the buffalo trails when they rode off on ranch chores for their father. They rode young and rode long, and often went without water. Even now, six decades later, David will pass up a cool spring, saying, "If I drink now, I'll be thirsty all day." To cut cedar fence posts, they went with a wagon to Green Mountain, near Crooks Gap—a round trip of two weeks. In early fall, each year, they spent ten days going back and forth to the Rattlesnake Hills for stove wood. They took two wagons—four horses pulling each wagon— and they filled them with limber pine. They used axes, a two-handled saw. Near home, they mined coal with their father—from the erosional wonderland they called Castle Gardens, where a horse-drawn scraper stripped the overburden and exposed the seams of coal. Their father was adept at corralling wild horses, a skill that called for a horse and rider who could outrun these closest rivals to the wind. He caught more than he kept, put his Flatiron brand on the best ones and sold the others. Some of them escaped. David remembers seeing one clear a seven-foot bar in the wild-horse corral and not so much as touch it. When he and Allan were in their early teens, his father sent them repping—representing Love Ranch in the general roundup—and they stayed in cow camp with other cowboys, and often enough their sougans included snow. When they were out on the range, they slept out on the range, never a night in a tent. This was not a choice. It was a family custom.

In the earlier stretch of his life when John Love had slept out for seven years, he would wrap himself in his sougans and finish the package with the spring hooks and D-rings that closed his henskin. During big gales and exceptional blizzards, he looked around for a dry wash and the crease of an overhanging cutbank. He gathered sage and built a long fire—a campfire with the dimensions of a cot. He cooked his beans and bacon, his mutton, his sourdough, his whatever. After dinner, he kicked the fire aside and spread out his bedroll. He opened his waterproof packet of books and read by kerosene lamp. Then he blew out the light and went to sleep on warm sand. His annual expenditures were seventy-five dollars. This was a man who wore a long bearskin coat fastened with bone pegs in loops of rope. This was a man who, oddly enough, carried with him on the range a huge black umbrella—his summer parasol. This was a man whose Uncle John Muir had invented a device that started a fire in the morning while the great outdoorsman stayed in bed. And now this wee bairn with the light-gold hair was, in effect, questioning Love Ranch policy by asking his father what he had against tents. "Laddie, you don't always have one available," his father said patiently. "You want to get used to living without it." Tents, he made clear, were for a class of people he referred to as "pilgrims."

When David was nine, he set up a trap line between the Hay Meadow and the Pinnacles (small sandstone buttes in Castle Gardens). He trapped coyotes, bobcats, badgers. He shot rabbits. He ran the line on foot, through late-autumn and early-winter snow. His father was with him one cold and blizzarding January day when David's rifle and the rabbits he was carrying slipped from his hands and fell to the snow. David picked up the gun and soon dropped it again. "It was a cardinal sin to drop a rifle," he says. "Snow and ice in the gun barrel could cause the gun to blow up when it was fired." Like holding on to a saddle horn, it was something you just did not do. It would not have crossed his father's mind that David was being careless. In sharp tones, his father said, "Laddie, leave the rabbits and rifle and run for home. Run!" He knew hypothermia when he saw it, no matter that it lacked a name.

Even in October, a blizzard could cover the house and make a tunnel of the front veranda. As winter progressed, rime grew on the nailheads of interior walls until white spikes projected some inches into the rooms. There were eleven rooms. His mother could tell the outside temperature by the movement of the frost. It climbed the nails about an inch for each degree below zero. Sometimes there was frost on nailheads fifty-five inches up the walls. The house was chinked with slaked lime, wood shavings, and cow manure. In the wild wind, snow came through the slightest crack, and the nickel disks on the dampers of the heat stove were constantly jingling. There came a sound of hooves in cold dry snow, of heavy bodies slamming against the walls, seeking heat. John Love insulated his boots with newspapers— as like as not *The New York Times*. To warm the boys in their beds on cold nights, their mother wrapped heated flatirons in copies of *The New York Times*. The family were subscribers. Sundays only. The *Times*, David Love recalls, was "precious." They used it to insulate the house: pasted it against the walls beside *The Des Moines Register, The Tacoma News Tribune*— any paper from anywhere, without fine distinction. With the same indiscriminate voracity, any paper from anywhere was first read and reread by every literate eye in every cow camp and sheep camp within tens of miles, read to shreds and passed along, in tattered circulation on the range. There was, as Love expresses it, "a starvation of print." Almost anybody's first question on encountering a neighbor was "Have you got any newspapers?"

The ranch steadings were more than a dozen buildings facing south, and most of them were secondhand. When a stage route that ran through the ranch was abandoned, in 1905, John Love went down the line

shopping for moribund towns. He bought Old Muskrat
—including the hotel, the post office, Joe Lacey's Muskrat Saloon—and moved the whole of it eighteen miles.
He bought Golden Lake and moved it thirty-three. He
arranged the buildings in a rough semicircle that embraced a corral so large and solidly constructed that other
ranchers travelled long distances to use it. Joe Lacey's
place became the hay house, the hotel became in part a
saddlery and cookhouse, and the other buildings, many
of them connected, became all or parts of the blacksmith shop, the chicken hatchery, the ice shed, the
buggy shed, the sod cellar, and the bunkhouse—social
center for all the workingmen from a great many miles
around. There was a granary made of gigantic cottonwood logs from the banks of the Wind River, thirty
miles away. There were wool-sack towers, and a wooden
windmill over a hand-dug well. The big house itself was
a widespread log collage of old town parts and original
construction. It had wings attached to wings. In the
windows were air bubbles in distorted glass. For its
twenty tiers of logs, John had journeyed a hundred miles
to the lodgepole-pine groves of the Wind River Range,
returning with ten logs at a time, each round trip requiring two weeks. He collected a hundred and fifty
logs. There were no toilets, of course, and the family had
to walk a hundred feet on a sometimes gumbo-slick path
to a four-hole structure built by a ranch hand, with
decorative panelling that matched the bookcases in the
house. The cabinetmaker was Peggy Dougherty, the
stagecoach driver who had first brought Miss Waxham
through Crooks Gap and into the Wind River country.
    The family grew weary of carrying water into the

house from the well under the windmill. And so, as she would write in later years:

After experiments using an earth auger and sand point, John triumphantly installed a pitcher pump in the kitchen, a sink, and drain pipe to a barrel, buried in the ground at some distance from the house. This was the best, the first, and at that time the only water system in an area the size of Rhode Island.

In the evenings, kerosene lamps threw subdued yellow light. Framed needlework on a wall said "WASH & BE CLEAN." Everyone bathed in the portable galvanized tub, children last. The more expensive galvanized tubs of that era had built-in seats, but the Loves could not afford the top of the line. On the plank floor were horsehide rugs—a gray, a pinto—and the pelt of a large wolf, and two soft bobcat rugs. Chairs were woven with rawhide or cane. John recorded the boys' height on a board nailed to the inside of the kitchen doorframe. A brass knocker on the front door was a replica of a gargoyle at Notre-Dame de Paris.

The family's main sitting and dining room was a restaurant from Old Muskrat. On the walls were polished buffalo horns mounted on shields. The central piece of furniture was a gambling table from Joe Lacey's Muskrat Saloon. It was a poker-and-roulette table— round, covered with felt. Still intact were the subtle flanges that had caused the roulette wheel to stop just where the operator wished it to. And if you reached in under the table in the right place you could feel the brass slots where the dealer kept wild cards that he

could call upon when the fiscal integrity of the house
was threatened. If you put your nose down on the felt,
you could almost smell the gunsmoke. At this table
David Love received his basic education—his schoolroom a restaurant, his desk a gaming table from a
saloon. His mother may have been trying to academize
the table when she covered it with a red-and-white
India print.

From time to time, other schoolmarms were provided by the district. They came for three months in
summer. One came for the better part of a year. By
and large, though, the boys were taught by their mother.
She had a rolltop desk, and Peggy Dougherty's glassed-
in bookcases. She had the 1911 Encyclopædia Britan-
nica, the Redpath Library, a hundred volumes of Greek
and Roman literature, Shakespeare, Dickens, Emerson,
Thoreau, Longfellow, Kipling, Twain. She taught her
sons French, Latin, and a bit of Greek. She read to them
from books in German, translating as she went along.
They read the Iliad and the Odyssey. The room was at
the west end of the ranch house and was brightly illu-
minated by the setting sun. When David as a child saw
sunbeams leaping off the books, he thought the contents
were escaping.

In some ways, there was more chaos in this remote
academic setting than there could ever be in a grade
school in the heart of a city.

The house might be full of men, waiting out a storm,
or riding on a round-up. I was baking, canning, washing
clothes, making soap. Allan and David stood by the gasoline
washing machine reading history or geography while I put

sheets through the wringer. I ironed. They did spelling be-side the ironing board, or while I kneaded bread; they gave the tables up to 15 times 15 to the treadle of the sewing machine. Mental problems, printed in figures on large cards, they solved while they raced across the . . . room to write the answers . . . and learned to think on their feet. Nine written problems done correctly, without help, meant no tenth problem. . . . It was surprising in how little time they finished their work—to watch the butchering, to help drive the bawling calves into the weaning pen, or to get to the corral, when they heard the hoofbeats of running horses and the cries of cowboys crossing the creek.

No amount of intellectual curiosity or academic discipline was ever going to hold a boy's attention if someone came in saying that the milk cow was mired in a bog hole or that old George was out by the wild-horse corral with the biggest coyote ever killed in the region, or if the door opened and, as David recalls an all too typical event, "they were carrying in a cowboy with guts ripped out by a saddle horn." The lessons stopped, the treadle stopped, and she sewed up the cowboy.

Across a short span of time, she had come a long way with these bunkhouse buckaroos. In her early years on the ranch, she had a lesser sense of fitting in than she would have had had she been a mare, a cow, or a ewe. She did not see another woman for as much as six months at a stretch, and if she happened to approach a group of working ranch hands they would loudly call out, "Church time!" She found "the sudden silence . . . appalling." Women were so rare in the country that when she lost a glove on the open range, at least twenty miles from home, a stranger who found it learned easily whose it must be and rode to the ranch to return it. Men did the housekeeping and the cooking, and went off to buy provisions at distant markets. Meals prepared in the bunkhouse were carried to a sheep wagon, where she and John lived while the big house was being built and otherwise assembled. The Wyoming sheep wagon was the ancestral Winnebago. It had a spring bed and a kitchenette.

After her two sons were born and became old enough to coin phrases, they called her Dainty Dish and sometimes Hooty the Owl. They renamed their food, calling it, for example, dog. They called other entrées caterpillar and coyote. The kitchen stool was Sam. They named a Christmas-tree ornament Hopping John. It had a talent for remaining unbroken. They assured each other that the cotton on the branches would not melt. David decided that he was a camel, but later changed his mind and insisted that he was "Mr. and Mrs. Booth." His mother described him as "a light-footed little elf." She noted his developing sense of scale when he said to her, "A coyote is the whole world to a flea."

One day, he asked her, "How long does a germ live?"

She answered, "A germ may become a grandfather in twenty minutes."

He said, "That's a long time to a germ, isn't it?"

She also made note that while David was the youngest person on the ranch he was nonetheless the most adroit at spotting arrowheads and chippings.

When David was five or six we began hunting arrowheads and chippings. While the rest of us labored along
scanning gulches and anthills, David rushed by chattering
and picking up arrowheads right and left. He told me once,
"There's a god of chippings that sends us anthills. He lives
in the sky and tinkers with the clouds."

The cowboys competed with Homer in the enter-
tainment of Allan and David. There was one who—as
David remembers him—"could do magic tricks with a
lariat rope, making it come alive all around his horse,
over our heads, under our feet, zipping it back and forth
around us as we jumped up and down and squealed
with delight." Sombre tableaux, such as butcherings,
were played out before them as well. Years later, David
would write in a letter:

We always watched the killing with horror and curiosity, although we were never permitted to participate at
that age. It seemed so sad and so irrevocable to see the
gushing blood when throats were cut, the desperate gasps
for breath through severed windpipes, the struggle for and
the rapid ebbing of life, the dimming and glazing of wide
terrified eyes. We realized and accepted the fact that this
was one of the procedures that were a part of our life on the
range and that other lives had to be sacrificed to feed us.
Throat-cutting, however, became a symbol of immediate
death in our young minds, the ultimate horror, so dreadful
that we tried not to use the word "throat."

He has written a recollection of the cowboys, no
less frank in its bequested fact, and quite evidently the
work of the son of his mother.

The cowboys and horse runners who drifted in to the ranch in ever-increasing numbers as the spring advanced were lean, very strong, hard-muscled, taciturn bachelors, nearly all in their twenties and early thirties. They had been born poor, had only rudimentary education, and accepted their lot without resentment. They worked days that knew no hour limitations but only daylight and dark, and weeks that had no holidays. . . . Most were homely, with prematurely lined faces but with lively eyes that missed little. None wore glasses; people with glasses went into other kinds of work. Many were already stooped from chronic saddle-weariness, bowlegged, hip-sprung, with unrepaired hernias that required trusses, and spinal injuries that required a "hanging pole" in the bunkhouse. This was a horizontal bar from which the cowboys would hang by their hands for 5-10 minutes to relieve pressure on ruptured spinal disks that came from too much bronc-fighting. Some wore eight-inch-wide heavy leather belts to keep their kidneys in place during prolonged hard rides.

When in a sense it was truly church time—when cowboys were badly injured and in need of help—they had long since learned where to go. David vividly remembers a moment in his education which was truncated when a cowboy rode up holding a bleeding hand. He had been roping a wild horse, and one of his fingers had become caught between the lariat and the saddle horn. The finger was still a part of his hand but was hanging by two tendons. His mother boiled water, sterilized a pair of surgical scissors, and scrubbed her hands and arms. With magisterial nonchalance, she "snipped the tendons, dropped the finger into the hot coals of the fire box, sewed a flap of skin over the stump,

[89]

smiled sweetly, and said: 'Joe, in a month you'll never know the difference.'"

There was a pack of ferocious wolfhounds in the country, kept by another flockmaster for the purpose of killing coyotes. The dogs seemed to relish killing rattle-snakes as well, shaking the life out of them until the festive serpents hung from the hounds' jaws like fet-tuccine. The ranch hand in charge of them said, "They ain't happy in the spring till they've been bit. They're used to it now, and their heads don't swell up no more." Human beings (on foot) who happened to encounter these dogs might have preferred to encounter the rattlesnakes instead. One summer afternoon, John Love was working on a woodpile when he saw two of the wolfhounds streaking down the creek in the direction of his sons, whose ages were maybe three and four. "Laddies! Run! Run to the house!" he shouted. "Here come the hounds!" The boys ran, reached the door just ahead of the dogs, and slammed it in their faces. Their mother was in the kitchen:

The hounds, not to be thwarted so easily, leaped to-gether furiously at the kitchen windows, high above the ground. They shattered the glass of the small panes, and tried to struggle through, their front feet catching over the inside ledge of the window frame, and their heads, with slavering mouths, reaching through the broken glass. I had only time to snatch a heavy iron frying pan from the stove and face them, beating at those clutching feet and snarling heads. The terrified boys cowered behind me. The window sashes held against the onslaught of the hounds, and my blows must have daunted them. They dropped back to the ground and raced away.

In the boys' vocabulary, the word "hound" joined the word "throat" in the deep shadows, and to this day when David sees a wolfhound there is a drop in the temperature of the center of his spine.

The milieu of Love Ranch was not all wind, snow, freezing cattle, and killer dogs. There were quiet, lyrical days on end under blue, unthreatening skies. There were the redwing blackbirds on the corral fence, and the scent of moss flowers in spring. In a light breeze, the windmill turned slowly beside the wide log house, which was edged with flowers in bloom. Sometimes there were teal on the creek—and goldeneyes, pintails, mallards. When the wild hay was ready for cutting, the harvest lasted a week.

John liked to have me ride with them for the last load. Sometimes I held the reins and called "Whoa, Dan!" while the men pitched up the hay. Then while the wagon swayed slowly back over the uneven road, I lay nestled deeply beside Allan and David in the fragrant hay. The billowy white clouds moving across the wide blue sky were close, so close, it seemed there was nothing else in the universe but clouds and hay.

When the hay house was not absolutely full, the boys cleared off the dance floor of Joe Lacey's Muskrat Saloon and strapped on their roller skates. Bizarre as it may seem, there was also a Love Ranch croquet ground. And in winter the boys clamped ice skates to their shoes and flew with the wind up the creek. Alternatively, they lay down on their sleds and propelled themselves swiftly over wind-cleared, wind-polished black ice, with an anchor pin from a coyote trap in each hand. Almost every evening, with their parents, they played mah-jongg.

One fall, their mother went to Riverton, sixty-five miles away, to await the birth of Phoebe. For her sons, eleven and twelve, she left behind a carefully prepared program of study. In the weeks that followed, they were in effect enrolled in a correspondence school run by their mother. They did their French, their spelling, their arithmetic lessons, put them in envelopes, rode fifteen miles to the post office and mailed them to her. She graded the lessons and sent them back—before and after the birth of the baby.

Her hair was the color of my wedding ring. On her cheek the fingers of one hand were outspread like a small, pink starfish.

From time to time, dust would appear on the horizon, behind a figure coming toward the ranch. The boys, in their curiosity, would climb a rooftop to watch and wait as the rider covered the intervening miles. Almost everyone who went through the region stopped at Love Ranch. It had not only the sizable bunkhouse and the most capacious horse corrals in a thousand square miles but also a spring of good water. Moreover, it had Scottish hospitality—not to mention the for-bidding distance to the nearest alternative cup of coffee. Soon after Mr. Love and Miss Waxham were married, Nathaniel Thomas, the Episcopal Bishop of Wyoming, came through in his Gospel Wagon, accompanied by his colleague the Reverend Theodore Sedgwick. Sedgwick later reported (in a publication called *The Spirit of Missions*):

> We saw a distant building. It meant water. At this lonely ranch, in the midst of a sandy desert, we found a young woman. Her husband had gone for the day over the range. Around her neck hung a gold chain with a Phi Beta Kappa key. She was a graduate of Wellesley College, and was now a Wyoming bride. She knew her Greek and Latin, and loved her horse on the care-free prairie.

The bishop said he was searching for "heathen," and he did not linger.

Fugitive criminals stopped at the ranch fairly often. They had to—in much the way that fugitive criminals in lonely country today will sooner or later have to stop at a filling station. A lone rider arrived at the ranch one day with a big cloud of dust on the horizon behind him. The dust might as well have formed in the air the letters of the word "posse." John Love knew the rider, knew that he was wanted for murder, and knew that throughout the country the consensus was that the victim had "needed killing." The murderer asked John Love to give him five dollars, and said he would leave his pocket watch as collateral. If his offer was refused, the man said, he would find a way to take the money. The watch was as honest as the day is long. When David does his field geology, he has it in his pocket.

People like that came along with such frequency that David's mother eventually assembled a chronicle called "Murderers I Have Known." She did not publish the manuscript, or even give it much private circulation, in her regard for the sensitivities of some of the first families of Wyoming. As David would one day comment, "they were nice men, family friends, who had put away people who needed killing, and she did not wish to offend them—so many of them were such decent people."

One of these was Bill Grace. Homesteader and cowboy, he was one of the most celebrated murderers in central Wyoming, and he had served time, but people generally disagreed with the judiciary and felt that Bill, in the acts for which he was convicted, had only been "doing his civic duty." At the height of his fame, he stopped at the ranch one afternoon and stayed for dinner. Although David and Allan were young boys, they knew exactly who he was, and in his presence were struck dumb with awe. As it happened, they had come upon and dispatched a rattlesnake that day—a big one, over five feet long. Their mother decided to serve it creamed on toast for dinner. She and their father sternly instructed David and Allan not to use the word "rattlesnake" at the table. They were to refer to it as chicken, since a possibility existed that Bill Grace might not be an eater of adequate sophistication to enjoy the truth. The excitement was too much for the boys. Despite the parental injunction, gradually their conversation at the table fished its way toward the snake. Casually—while the meal was going down—the boys raised the subject of poisonous vipers, gave their estimates of the contents of local dens, told stories of snake encounters, and so forth. Finally, one of them remarked on how very good rattlers were to eat.

Bill Grace said, "By God, if anybody ever gave me rattlesnake meat I'd kill them."

The boys went into a state of catatonic paralysis. In the pure silence, their mother said, "More chicken, Bill?"

"Don't mind if I do," said Bill Grace.

ᏬᏌ Muskrat Creek was the second homestead on which John Love had filed in Wyoming. The first—thirty miles away—was in the Big Sand Draw, where the grass was inadequate, the snows were exceptionally deep, and the water was marginally potable. In 1897, he collapsed his umbrella and moved. At Muskrat Creek, long before he bought the stagecoach towns, he lived in an earth dugout roofed with pine poles and clay. It was warm in winter, cool in summer, and danker than Scotland all year round. He was prepared to run risks. In Lander, sixty miles west, he made an extraordinary bet with a bank, whose assets included a number of thousands of sheep. John Love bet that he could take them for a summer and return them in the fall, fatter on the average by at least ten pounds. If he succeeded, he would be paid handsomely. If he failed, he would receive a scant wage. He was taking a chance on the weather, because a bad storm could wipe out the flock. By November, the sheep were as round as poker chips, ready to be cashed in. Leaving them in the care of a herder, he rode to Thermopolis, where he made a down payment on a flock of his own. The conditions of the deal were rigid: the rest of the money was to be paid in seven days or the deposit was forfeit and the animals, too. Within the week, he would have to return to his fattened sheep, move them to Lander, collect his money, and return to Thermopolis—a round trip of two hundred and fifty miles. The sky over Thermopolis was dark with snowcloud. In his bearskin cap, his bearskin coat, his fleece-lined leather chaps, he saddled up Big Red— Big Red, whose life had begun somewhere in the Red Desert in 1888, a wild horse. The blizzard began as horse and rider were climbing the Owl Creek Mountains. Through steep terrain that would have been hazardous in warm clear weather, they felt their way in white-outs and darkness, in wind-chill factors greater than fifty below zero. Covering about six miles an hour, they reached the herd in twenty-one hours, and almost immediately began the gingerly walk to Lander, conserving the animals' weight. John won his bet, got back on Big Red, and flew across the mountains with the money. He and the horse beat the deadline. He collected his ewes, took them home, and bred them. In seven days, he had, among other things, set himself forward one year. By 1910, when he married Miss Waxham, he owned more than eleven thousand sheep and hundreds of cattle and horses—a fortune in livestock which today would be valued at roughly five million dollars.

In the early days of his marriage, John Love used to ride around his place reciting the verse of William Cowper:

I am monarch of all I survey,
My right there is none to dispute.

As he built up his new home, he did not seem worried that in recent years herders had been killed, wagons had been burned, and sheep had been clubbed to death or driven over cliffs by the thousand. As anyone who has seen three Western movies cannot help but know, there was bloody warfare between cattlemen and sheepmen; and well into the new century the strife continued. According to David, his father stocked the ranch with both cattle and sheep specifically as a way of getting along with both sides. His monarchy would be disputed only by nature and bankers.

Cowboys, meanwhile, made unlikely paperhangers.

Rolls of green figured wallpaper had arrived from a mail-order company. What to do with them, no one quite knew, but there were directions. I made dishpans full of paste. In the evening John called in the half dozen cowboys from the bunkhouse. They carried planks and benches. They put all the leaves in the wobbly dinner table. I measured and cut, pasted and trimmed lengths of wallpaper. Then in chaps and jingling spurs the cowpunchers strode along the benches, slapping paste brushes and dangling strips of torn wallpaper over the dining room ceiling. We were all surprised and tremendously pleased with the results and celebrated over a ten-gallon keg of cider.

John put a roof on the ranch house that was half clay and a foot thick. It consisted of hundreds of two-inch poles covered with burlap covered with canvas covered with rafters embedded in the clay, with corrugated iron above that, coated with black asphaltum. It helped the house be cool in summer, warm in winter—and in the Wind River Basin was unique. But while this durable roof could defend against Wyoming weather the rest of the ranch could not. In the winter of 1912, winds with velocities up to a hundred miles an hour caused sheep to seek haven in dry gulches, where snows soon buried them as if in avalanche. Going without sleep for forty and fifty hours, John Love and his ranch hands struggled to rescue them. They dug some out, but many thousands died. Even on the milder days, when the temperature came up near zero, sheep could not penetrate the wind-crusted drifts and get at the grass below. The crust cut into their legs. Their tracks were reddened with blood. Cattle, lacking the brains even to imagine buried grass, ate their own value in cottonseed cake. John Love had to borrow from his bankers in Lander to pay his ranch hands and buy supplies.

That spring, a flood such as no one remembered all but destroyed the ranch. The Loves fled into the night, carrying their baby, Allan.

At daylight we returned to the house. Stench, wreckage and debris met us. The flood had gone. Its force had burst open the front door and swept a tub full of rainwater into the dining room. Chairs and other furniture were overturned in deep mud. Mattresses had floated. Doors and drawers were already too much swollen for us to open or shut. The large wardrobe trunk of baby clothes was upset. Everything in it was soaked and stained. Around all the rooms at the height of the tabletops was a water mark, fringed with dirt, on the new wallpaper.

Almost immediately, the bankers arrived from Lander. They stayed for several amiable days, looked over the herd tallies, counted surviving animals, checked John Love's accounts. Then, at dinner one evening, the bank's vice-president rubbed his hands together and said to his valued customer, his trusted borrower, his first-name-basis longtime friend, "Mr. Love, we need more collateral." The banker also said that while John Love was a reliable debtor, other ranchers were not, and others' losses were even greater than Love's. The bank, to protect its depositors, had to use Love Ranch to cover itself generally. "We are obliged to cash in on your sheep," the man went on. "We will let you keep your cattle—on one condition." The condition was a mortgage on the ranch. They were asking for an interest in the land of a homesteader who had proved up.

John Love shouted, "I'll have that land when your bones are rotting in the grave!" And he asked the man to step outside, where he could curse him. To the banker's credit, he got up and went out to be cursed. Buyers came over the hill as if on cue. All surviving sheep were taken, all surviving cattle, all horses—even dogs. The sheep wagons went, and a large amount of equipment and supplies. John Love paid the men in the bunkhouse, and they left. As his wife watched the finish of this scene, standing silent with Allan in her arms, the banker turned to her kindly and said, "What will you do with the baby?"

She said, "I think I'll keep him."

It was into this situation that John David Love was born—a family that had lost almost everything but itself, yet was not about to lose that. Slowly, his father

assembled more modest cavvies and herds, beginning
with the capture of wild horses in flat-out all-day rides,
maneuvering them in ever tighter circles until they were
beguiled into entering the wild-horse corral or—a few
miles away—the natural cul-de-sac (a small box canyon)
known to the family as the Corral Draw. Watching one
day from the granary roof, the boys—four and five—in
one moment saw their father on horseback crossing the
terrain like the shadow of a cloud and in the next saw
his body smash the ground. The horse had stepped in
a badger hole. The rider—limp and full of greasewood
punctures, covered with blood and grit—was uncon-
scious and appeared to be dead. He was carried into the
house. After some hours, he began to stir, and through
his pain mumbled, "That damned horse. That damned
horse—I never did trust him." It was the only time in
their lives that his sons would hear him swear.

There were periods of drought, and more floods,
and long, killing winters, but John Love never sold out.
He contracted and survived Rocky Mountain spotted
fever. One year, after he shipped cattle to Omaha he
got back a bill for twenty-seven dollars, the amount by
which the cost of shipment exceeded the sale price of
the cattle. One spring, after a winter that killed many
sheep, the boys and their father plucked good wool off
the bloated and stinking corpses, sold the wool, and
deposited the money in a bank in Shoshoni, where the
words "Strength," "Safety," and "Security" made an
arc above the door. The bank failed, and they lost the
money. Of many bad winters, the worst began in 1919.
Both David and his father nearly died of Spanish in-

fluenza, and were slow to recuperate, spending months in bed. There were no ranch hands. At the point when the patients seemed most in danger, his mother in her desperation decided to try to have them moved to a hospital (a hundred miles away), and prepared to ride for help. She had the Hobson's choice of a large, rebellious horse. She stood on a bench and tried to harness him. He kicked the bench from under her, and stepped on her feet. She gave up her plan.

The bull broke into the high granary. Our only, and small, supply of horse and chicken feed was there. Foolishly, I went in after him and drove him out down the step. Cows began to die, one here, one there. Every morning some were unable to rise. By day, one walking would fall suddenly, as if it had no more life than a paper animal, blown over by a gust of wind.

The bull actually charged her in the granary and came close to crushing her against the back wall. She confused it, sweeping its eyes with a broom. It would probably have killed her, though, had it not stepped on a weak plank, which snapped. The animal panicked and turned for the door. (In decades to follow, John Love never fixed the plank.)

Snow hissed around the buildings, wind blew some snow into every room of the closed house, down the chimney, between window sashes, even in a straight shaft through a keyhole. The wood pile was buried in snow. The small heap of coal was frozen into an almost solid chunk of coal and ice. In the numbing cold, it took me five hours a day to bring in fuel, to carry water and feed to the chickens, to put out hay and cottonseed cake for the cattle and horses.

. . .

John began to complain, a favorable sign. Why was I outside so much? Why didn't I stay with him? To try to make up to him for being gone so long, I sat on the bed at night, wrapped in a blanket, reading to him by lamplight.

Somewhere among her possessions was a letter written to her by a Wellesley friend asking, "What do you do with your spare time?"

Where the stage route from Casper to Fort Washakie had crossed a tributary of Muskrat Creek, the banks were so high and the drop to the creekbed so precipitous that the site was littered with split wagon reaches and broken wheels. Allan and David called it Jumping-Off Draw, its name on the map today. Finding numerous large bones in a meadowy bog, they named the place Buffalo Wallows. Indians had apparently driven the bison into the swamp to kill them. One could infer that. One could also see that the swamp was there because water was bleeding from rock outcrops above the meadows. In a youth spent on horseback, there was not a lot to do but look at the landscape. The rock that was bleeding water was not just porous but permeable. It was also strong. It was the same red rock that the granary stood on, and the bunkhouse. Very evidently, it was made of naturally cemented sand. The water could not have come from the creek. The Buffalo Wallows were sixty feet higher than the creek. The sandstone layers tilted north. They therefore reached out to the east and west. There was high ground to the east. The water must be coming down from there. One did not need a Ph.D. from Yale to figure that out—especially if one was growing up in a place where so much rock was exposed. Pending further study, his interpretation of the Buffalo Wallows was just a horseback guess. All through his life, when he would make a shrewd surmise he would call it a horseback guess.

The water in the sandstone produced not only the bogs but the adjacent meadows as well—in this otherwise desiccated terrain. From the meadows came hay. There was an obvious and close relationship between bedrock geology and ranching. David would not have articulated that in just those words, of course, but he thought about the subject much of the time, and he was drawn to be a geologist in much the way that someone growing up in Gloucester, Massachusetts, would be drawn to be a fisherman. "It was something to think about on long rides day after day when everything was so monotonous," he remarked not long ago. "Monotony was what we fought out there. Day after day, you had nothing but the terrain around you—you had nothing to think about but why the shale had stripes on it, why the boggy places were boggy, why the vegetation grew where it did, why trees grew only on certain types of rock, why water was good in some places and bad in others, why the meadows were where they were, why some creek crossings were so sandy they were all but impassable. These things were very real, very practical. If you're in bedrock, caliche, or gumbo, the going is hard. Caliche is lime precipitate at the water table—you learn some geology the hard way. There was nothing else to be interested in. Everything depended on geology. Any damn fool could see that the vegetation was directly responsive to the bedrock. Hence birds and wildlife were responsive to it. We were responsive to it. In winter, our life was governed by where the wind blew, where snow accumulated. We could see that these natural phenomena were not random—that they were controlled, that there was a system. The processes of erosion and deposition were things we grew up with. An insulated society does not see how important terrain is to someone who has to understand it in order to live with it. Much of it meant life or death for the animals, and therefore survival for us. If there was one thing we learned, it was that you don't fight nature. You live with it. And you make the accommodations—because nature does not accommodate."

In the driest months, he saw mud cracks so firm a horse could step on them without breaking their polygonal form. When he saw the same patterns in rock, he had no difficulty discerning that the rock had once been mud and that the cracks within it were the preserved summer of a former world. In the Chalk Hills (multicolored badlands), getting down from his mount, he found the tiny jaws and small black teeth of what he eventually learned were Eocene horses—the first horses on earth, four hands tall.

Among the figures that appeared on the horizon and slowly approached the ranch—and sometimes stayed indefinitely—were geologists. The first he met were from the United States Geological Survey. Others worked for oil companies. The oilmen were well dressed and had shiny boots that caught his eye. Some of these people were famous in the science—for example,

Charles T. Lupton, a structural geologist who had located the wildcats of the Cat Creek Anticline and discovered the oil of Montana. He did something like it in the Bighorn Basin. David particularly remembers him on two counts: first, that he "talked about the outside world," and, second, that he came in off the range with fragments of huge ammonites—index fossils of the Upper Cretaceous—and demonstrated by extrapolation that these spiral cephalopods had approached the size of wagon wheels. Lupton's obituary in the *Bulletin of the American Association of Petroleum Geologists* says, "Always he had a word to say to the children of his friends."

That was written in 1936, by Charles J. Hares, who had also made frequent stops at Love Ranch. Hares (1881–1970), in the course of a career in the Geological Survey and private business, became "the dean of Rocky Mountain petroleum geologists" and was one of the founders of the Wyoming Geological Association. His work on the anticlines of central Wyoming set up most of the major oil discoveries in the region. He was a celebrated teacher as well, and his roster of youthful field assistants in time became a list of some of the most accomplished geologists in America. Geologists who came to the ranch were reconnaissance geologists of the first rank, who went into unknown country and mapped it with an accuracy that is remarkable to this day. In David's words: "They raised a magic curtain. They showed us things we'd never seen. There was mother-of-pearl on some of the ammonites. There were Mesozoic oyster shells with both valves intact. You could open them up and see inside. All these things were marine—

known only from ocean floors. They also brought in beautiful leaves, fifty million years old, from non-marine rocks of the Eocene. The seas were gone. The mountains had come up. Day after day, we could look around us and see, in the mind's eye, those things happening."

David's mother owned Joseph LeConte's "Elements of Geology." He read it when he was nine years old. Did he grasp structure and stratigraphy then? Could he have begun to understand faulting? "To some extent, yes," he says. "After all, we could see it out in front of us."

On the southern horizon were the Gas Hills—a line of blue-banded ridges formed in a wedge like the prow of a ship (actually, an arch of shale). David would find uranium there in 1953. Riding over those ridges as a boy, he smelled gas. There were oil seeps as well. ("It was something you could relate to. The Gas Hills weren't called that for nothing.")

Oil and gas had entered the conversation at the ranch when David was four years old. In that summer (1917), derricks suddenly appeared in six different places within twenty miles; and, like other ranchers, the Loves began to muse upon a solvency giddily transcending the wool of frozen sheep. David's mother referred to all this as the family mirage.

Oil to us was once just a word recurring through the story of Wyoming. Indians and trappers told of curious oil seeps. Captain Bonneville in 1832 wrote about finding the "great tar springs" near what is now Lander. His party used the oil as a remedy for the cracked hoofs and harness sores of their horses, and as a "balsam" for their own aches and pains. Jim Bridger, scout, Indian fighter, and fort builder, mixed tar with flour and sold it along the Oregon Trail to emigrants, who needed axle grease for their wagons. They found, too, that buffalo chips made a hotter fire when a little tar was added to them.

And now she told the visiting geologists that if oil was what they were looking for they would surely find it under the ranch, because her younger son's initials were J.D.

Such excitement was contagious. Into our repetitive talk of sheep, cattle, horses, weather, and markets, new words appeared: anticline, syncline, red beds, sump, casing, drill stem, bits, crow's nest, cat walk, headache beam. Almost every herder had his own oil dome. We took up oil claims.

• • •

A range ne'er-do-well, grizzled and tattered, caught a ride to our house. He inquired importantly whether he might stay with us a few days while he did some validating work on his oil claims. Then he asked John if he might borrow a shovel. But to get to his claims, he said, he needed a team and a wagon. Having succeeded so far, he demanded, "Now, where's your oil?"

The boys might be far from sidewalks, but they would not grow up naïve.

A man named Jim Roush had a way of finding oil without a drill bit. Arriving at the ranch, he offered his services. Jim Roush was sort of a Music Man—an itinerant alchemist of structure, a hydrocarbonic dow-

ser. He had a bottle that was wrapped in black friction
tape. It dangled from a cord, and contained a secret
fluid tomographically syndetic with oil. While David
Love looked on—with his brother, his mother, and his
father—Roush stood a few feet from their house and
suspended the bottle, which began to spin. Light flashed
on one hand—from a large apparent diamond. In silent
concentration, he counted. Ypresian, Albian, Hauteri-
vian, Valanginian—there was a geologic age in every
spin. When the bottle stopped, its aggregate revolutions
could be factored as depth to oil. David never saw Jim
Roush again, except to the extent that his ghost might
haunt the Geological Survey.

In 1918, a hundred-million-dollar oil-and-gas field
was discovered at Big Sand Draw, where John Love
gave up his first homestead, in 1897. After the Mineral
Lands Leasing Act of 1920, oil companies could obtain
leases directly from the government. A rancher's claims
no longer intervened. A rancher needed fifty thousand
dollars to drill on his own.

It was the general opinion on the range that if a man
had that kind of money he did not need an oil well. Our
mirage disappeared completely.

Emblematically, fire broke out in the oil fields of
Lost Soldier, fifty miles southwest, and for weeks
lighted up the night sky. In Horseshoe Gulch, six miles
from the ranch, Sinclair Wyoming drilled forty-three
hundred feet and found gas, which came out with such
force that it destroyed the drill stem and blew the
wooden derrick to pieces.

When the blast came, the driller was carrying a hundred-pound anvil across the rig floor. He told us that he raced half a mile over the sagebrush before he realized that he still held the anvil.

The Loves hitched up a wagon and went after the wood. They would burn the entire derrick in their kitchen stove.

David picked up a small, rough chunk of soft gray shale, blasted out of the depths of the well. He saw in it tiny marine fossils and fragile, lustrous pieces of mother-of-pearl, the size of his fingernail, that had once held the bodies of living clams; they came from more than half a mile underground; they lived before the time of men on the earth; they had been buried, how many ages, since they moved about that unseen shore. The driller told him that those shells predicted the presence of oil. . . . He brought home the rock with the delicate shells embedded in it and has kept it ever since in an Indian bowl.

When the boys were teen-aged, they occasionally saddled up and rode twenty-six miles to dances in Shoshoni. (I once asked David if they were square dances, and he said, "No. It was contact sport.") After dancing half the night, they rode twenty-six miles home. Their mother rented a house in Lander and stayed there with them while they attended Fremont County Vocational High School. One of their classmates was William Shakespeare, whose other name was War Bonnet. Lander at that time was the remotest town in Wyoming. It advertised itself as "the end of the rails and the start of the trails." Now and again, when the boys and their mother went visiting, they went through Red Canyon.

A scene of great beauty, long sinuous Red Canyon was a presentation of the Mesozoic, framed in wide margins of time. In the eastern escarpment, the rock, tilting upward, protruded from the earth in cliffs six hundred feet high, and these were Eocene over Paleocene over Cretaceous and Jurassic benches above the red Triassic wall. Upward to the west ran a sage-covered Permian slope, on a line of sight that led higher in altitude and lower in time to the Precambrian roof of the Wind River Range—peaks on the western horizon.

Within the Triassic red was a distinctive white line that ran on as far as the eye could see. It was amazingly consistent, five feet thick, a limestone. In time, David would learn that this uniform bed covered fifty thousand square miles and was one of the most unusual rock units anywhere in the world, with such an absence of diagnostic fossils that no one could tell if the water it formed in was fresh or salt. ("It is a major marker bed of the Rockies," he once said, pointing it out to me. "Fifty thousand square miles—try to imagine any place in the world today where you can find that kind of stability. I can't. It's unique geologically.") As a youth, though, he was less fascinated by this Alcova limestone than by other aspects of Red Canyon. A woman who lived there was known as Red Canyon Red, for her striking Triassic hair. She was—as he would in later years describe her—"a whore *par excellence*." This may have been one reason that the boys' mother routinely accompanied her sons when they went through Red Canyon Red's canyon.

To shorten the trip to Lander, and make his own visits more frequent, John Love bought a used Buick.

Under severe nervous and vocal strain, he taught himself to drive, alone, on a wide expanse of level ground. Automatically he called "Whoa, there!" when he wanted to stop.

He knew a stockman who, in a similar effort, had failed, and had destroyed his car with an axe. John was resolved not to let that happen to him. He triumphed, of course, and the family was soon cruising to the Sweetwater Divide, with a picnic lunch and a jug of lemonade. Their horizons, already wide, before long rapidly expanded as John decided to take the first vacation of his life in order that the boys might see the Pacific Ocean before they went to college. The Loves headed west in the Buick. They had a breadbox, a camp stove, a nest of aluminum pots. The back seat was stacked high with blankets. Suitcases rode on the running boards, and on top of the luggage was a tepee.

The boys went to the University of Wyoming, where Allan majored in civil engineering. David majored in geology, and was elected to Phi Beta Kappa. Words carved into the university sandstone said:

> STRIVE ON—THE
> CONTROL OF NATURE
> IS WON NOT GIVEN

David stayed in Laramie to earn a master's degree, and later, on a scholarship, went east to seek a Ph.D. at Yale. Arriving with some bewilderment in that awesome human topography, he noticed a line from Rafael Sabatini carved in stone in a courtyard of the Hall of Graduate Studies: "He was born with a gift for laughter and a sense that the world is mad." Those words steadied him at Yale, and helped prepare him for a lifetime in government and science. As a graduate student, he had to advance his reading knowledge of German, which he did over campfires on summer field work in the mountains of Wyoming. One book mentioned an inscription above a doorway at the German Naval Officers School, in Kiel—an unlikely place for a Rocky Mountain geologist to discover what became for him a lifelong professional axiom. As he renders it in English: "Say not 'This is the truth' but 'So it seems to me to be as I now see the things I think I see.'"

Yale had one of the better geology departments in the world, and its interests were commensurately global. It was syllogistic, encyclopedic, and stirred its students to extended effort—causing him to disappear into the library for months on end in what he calls his golden years. It was a department preoccupied with the Big Picture, and as a result it was not overcrowded with people who had seen a lot of outcrops. That, at any rate, is how it seemed to a student who had seen almost nothing but outcrops—close at hand, or slowly turning from the perspective of a saddle. In no way did this distinction diminish the reverence he felt for these Eastern petrologues. "Their field geology was, let's say, incomplete," he will remark tenderly.

He did his field work in exceptionally rugged country—in the Tetons for a time, during those grad-school summers, but mainly along the southern margin of the Absaroka Range, roughly a hundred miles north-west of the ranch. He chose an area of about three hundred thousand acres (five hundred square miles) with intent to develop an understanding of it sufficient for the completion of a doctoral thesis. Geologically, it was a blank piece of the earth. Virtually nothing was known. The area had been mapped topographically. He took the map with him. Some of its streams ran up-hill.

The Absarokas, it seemed, were a multilayered pile of pyroclastic debris—sedimentary rock whose components had once been volcanic outpourings. It was material that—after hardening—had been crumbled by weather and collected and moved by streams. The Absaroka volcanic sediments were a local part of the vast fill that had buried the central Rocky Mountains—a hard and therefore durable part. Their huge boulders indicated close proximity to the vents from which the rock had poured. (On a relief map, the Absarokas seem to spill out of Yellowstone Park.) During the Exhumation of the Rockies, the durability of these formations had left them in place. They resemble a battlement, standing seven thousand feet above the adjacent plains.

When Love chose this thesis topic, he was not choosing a journey from A to B. He did not mark off a little basin somewhere and essay to describe the porridge it contained. He picked a spinning pinwheel of geology, with highlights flashing from every vane. For example, the western end of the Owl Creek Mountains

[113]

is still buried—under the Absarokas in the area of Love's thesis. Another chain of mountains is largely buried there, too. He discovered it and named it the Washakie Range. That the Absarokas were structurally separate from the Owl Creeks was obvious. It was not so readily apparent that the Owl Creeks were younger than the Washakies, or that the Washakies in their early days had been thrust southwestward over undistorted shales on the floor of the Wind River Basin. Gradually, he figured these things out, alone in the country, on foot. His thesis area reached a short distance into the Wind River Basin, and thus completed in its varied elements the panoply of the Rockies. It included folded mountains and dissected plateaus. It included basin sediments and alpine peaks, dry gulches and superposed master streams, desert sageland and evergreen forest rising to a timberline at ten thousand four hundred feet. It contained the story unabridged—from the preserved sub-summit surface to the fossil topographies exhumed far below. At any given place in the area, temperatures could change eighty degrees in less than a day. The territory was roadless, and after a couple of years on foot he was ready to bring in a horse. Methodically cross-referencing the lithology, the paleontology, the stratigraphy, the structure, he mapped the region geologically—discovering and naming seven formations.

In one of those Yale summers, while taking some time away from the rock, he badly cut his foot in a lake near Lander. He made a tourniquet with his bandanna, and limped into town to see Doc Smith. This was Francis Smith, M.D., who had coaxed David's father past the tick fever, had seen David's mother through a strep infection that nearly killed her, and, over the years, had put enough stitches in David to complete a baseball. Now, as he worked on the foot, he told David that one of his recent office visitors had been Robert LeRoy Parker himself (Butch Cassidy).

David said politely that Cassidy was dead in Bolivia, and everybody knew that.

Smith said everybody was wrong. The patient had appeared in the doorway, and had stood there long and thoughtfully, searching the face of the doctor. Pleased by what he did not find, he said. "You don't know who I am, do you?"

The doctor said, "You look familiar, but I can't quite say."

The patient remarked that his face had been altered by a surgeon in Paris. Then he lifted his shirt, exposing the deep crease of a repaired bullet wound—craftsmanship that Doc Smith recognized precisely as his own.

The work that David Love was doing in Wyoming attracted the attention of the Geological Society of America. He was invited to speak at the society's national meeting, which, in the course of its migrations, happened to be scheduled for Washington, D.C. He suffered great anticipatory fright. It was most unusual for a graduate student to be asked to speak to the G.S.A. He was intimidated—by the East in general, by the capital city, by the fact that the foremost geologists in America would be there. Bailey Willis, of Stanford, would be there; Andrew Lawson, of Berkeley; Walter Granger, of the American Museum of Natural History; Taylor Thom, of Princeton. The paper that Love presented was on folding and faulting in Tertiary rocks. The Tertiary period runs from sixty-five million to something like two million years before the present. When Love went to Yale, the conventional wisdom in geology held that all folded and faulted rock was older than the Tertiary—that all Tertiary rock was undeformed. For his thesis in the Absarokas, he had mapped many areas of folded and faulted Tertiary rock. He knew the fossils, the stratigraphy. This was in no sense a horseback guess. He practiced carefully what he would say, and, when his moment came, there he was on a platform in the ballroom of the Hotel Washington struggling to control his voice, unaware that he had forgotten to button his suspenders. They were hanging down in back, exposed and flapping. His embarrassment had scarcely begun. At the climax of his presentation—as he described the deformation that had made clear to him that fifteen million years into Tertiary time the Laramide Revolution had not quite ended—he heard what he describes as hoots of derision, and when he finished there was no applause. The big room was silent. A moment passed, and then the structural geologist Taylor Thom, some of whose work was challenged by Love's paper, stood up and said, "This paper is a milestone in Rocky Mountain geology."

At his first G.S.A. meeting, a couple of years earlier in New York, about the only person he knew was Samuel H. Knight, his distinguished tutor from the University of Wyoming. Gradually, the faces and forms of strangers attached themselves to names long familiar to him on scientific papers and the spines of books. His personal pantheon came alive around him, and he was pleased to discover how easily approachable they were. When he tells of the experience—now that he is in his early seventies and the Grand Old Man of Rocky Mountain Geology—he could be describing himself: "They put their pants on one leg at a time. They were very human individuals. They encouraged young people to speak out." As they discussed one another's papers, he relished their candor, their style of disagreement. A paleontologist named Asa Mathews got up and presented what he believed was the discovery that the world's first bird had come into existence in Permian time—roughly a hundred million years earlier than Archaeopteryx, which reigned then (and does now) as first bird. Mathews detailed some remarkable trace fossils in Permian rock in Utah, which sequentially recorded, he said, a bird as it ran along the ground, its wingtips awkwardly scraping until, finally, it took off. Walter Granger arose before the assembly to greet this unusual news. He said, "Professor Mathews has undoubtedly demonstrated that the first bird flew over that part of Utah, but he has not demonstrated that it landed."

David Griggs, who was not much older than Love, gave a superb demonstration of some fresh ideas about mountain-building. Afterward, Bailey Willis, whom Love describes as "one of the grandfathers of structural geology in the world," anointed Griggs with praise. The great Andrew Lawson, who named the San Andreas Fault, was on his feet next, and virtually conferred upon Griggs an honorary degree by saying, "For the first and only time in my life, I agree with Bailey Willis."

In another context, a young geologist challenged Walter Granger, saying, "Dr. Granger, are you sure you're right?" Granger answered, without a flicker of hesitation, "Young man, I will consider myself a great success in life if I prove to be right fifty per cent of the time."

After Yale, Love worked a season for the U.S. Geological Survey in the Wasatch Mountains of Utah, scratching his way a step at a time through the dense, stiff branches of scrub oak that—ninety years before— had held up the Donner party enough to set the schedule for its eventual rendezvous with snow. Employed by the Shell Oil Company, he spent five years looking for areas of possible oil accumulation in the structures of Illinois, Indiana, Missouri, Kentucky, Georgia, Arkansas, Michigan, Alabama, and Tennessee. This increased his experience in ways that included a good deal more than rocks—especially in the southern Appalachians. Looking for outcrops, he walked many hundreds of miles of streambeds, "brushing water moccasins to one side." He studied roadcuts and railroad cuts. He slept where the day ended. Sometimes he stayed, for weeks at a time, with farmers. A farmer in Tennessee took him aside one day and offered him his unusually beautiful teen-age daughter in marriage if David would buy her a pair of shoes—her first. Apparently, the girl had developed such a longing for the young geologist that her father wanted her to have him. David felt that he was receiving "the ultimate compliment," for the farmer wanted to give him what the farmer valued most in the world. Earnestly, he wished not to offend, lest, among other things, he lose his advantage in a complicated country dissected by entrenched streams, a prime piece of the Mississippi Embayment, as geologists call the

great bulb of sediments that reaches from the mouth of
the Mississippi River as far north as Paducah, Kentucky.
The farmer was in a strategic position to permit the
young geologist to find certain permeable sandstones
wedging out between layers of shale in updips where
migrating petroleum might have become trapped. So he
was anxious not to ruffle the feelings of his host. Besides,
he was engaged.

In Laramie in 1934, David had met a geology student from Bryn Mawr College who had come west to
spend a couple of semesters at the University of Wyoming under an arrangement that he would ever after
refer to as her junior year abroad. Her father, in grant-
ing her permission to go to Wyoming, had commented
that everyone has a right—at least once in a lifetime—
to run away to sea. Her name was Jane Matteson, and
she had grown up in the quiet streets and private educational enclaves of Providence, Rhode Island. She had
twice the sophistication of this ranch hand, notwithstanding the fact that she wrote "crick" in her early
field notes, believing that Western geologists had taught
her a new term. Moreover, she considered him "too
good-looking." (Had he ever been inclined to, he could
have answered the complaint in kind.) He appealed
to her, though—in part because he kept his distance.
She liked her cowboys unaggressive, and this one (at
the time) was so shy—so reserved and respectful—that
he stayed on his own side of his little Ford coupé. Her
philosophy of conjugal evolution was contained in the
phrase "A kiss is a promise," and for the time being
there were no promises. He took her to the top of the
Laramie Range, and up there on the pink granite under

the luminous constellations gave her some gallant, if cryptic, advice. He said, "Whatever you do, don't come up here to look at the stars with a geologist."

When a letter arrived containing his acceptance at Yale, they went over the mountains to celebrate in Cheyenne. On the return trip, on old U.S. 30, they were met at the gangplank by a spring storm, and they worried that they would be snowed in for the night, with resulting damage to their reputations. Forging on through the blizzard, they made it to Laramie. After Jane finished Bryn Mawr, she did graduate work at Smith, returning to Wyoming for the summer field work in the Black Hills and the Bighorn Mountains which led to her master's thesis on Pennsylvanian-Permian rock. More apart than together while he completed his work at Yale, they exchanged long and frequent letters, which was her idea of a way for two people to get to know each other well and review their approaches to life. (When she was telling me this, not long ago, she added, with a bit of gemflash in her dark eyes, "I suppose that's why young people live together now— instead of writing letters.") In the nineteen-seventies, she edited a book on Mesozoic mammals for the University of California Press, but her own work in geology has been at best sporadic. "In our generation," she once remarked to me, "a woman's place was in the home raising a family if that was what she chose to do. I brooded a little bit about this but not much. You get to be twenty-five. You kick around the options. You decide you would rather be a wife and mother than a geologist. The fact that I can talk geology with him is just gravy." She offers her thoughts as an advocate of the geological devil to assist in the refinement of David's ideas. "What makes you think you know that? How could you possibly infer that?" she will say to him, not always stopping short of his Celtic irascibilities.

At Love Ranch, in 1910 or so, and apropos of who knows what, David's mother asked his father, "Is it true that it is necessary to kill a Scot or agree with him?"

John Love took the question seriously. After thinking it over, he answered, without elaboration, "No."

"David is not afraid of a new idea," Jane continues. "He's a pragmatist. He never looks back. He is both ingenious and practical. On the ranch when he grew up there was no plumbing, no electricity, no automobile— and the equipment they had they repaired themselves. If he has a piece of baling wire, he can fix anything. He fixes everything from plumbing to cars. He applies the same practicality to geology. If a slide block suggests that it might go downhill, he has the physics and he knows if it will work. His talent lies particularly in his sense of cause and effect. His knowledge, experience, and curiosity extend far beyond the mere presence of a rock. He is the most creative geologist I know."

They were married in 1940. Two of their four children were conceived after a summer field season and born during the next one, with him off in the wild, hundreds of miles away. That, says Jane, is just one more facet of geology. On a field trip somewhere in the late nineteen-seventies, David said to me, "I've been hearing about it for thirty-four years." The summer field season begins in June and ends about four months later, during which time she seldom saw much of him in their early years, a condition she regards as follows: "My father was a lawyer in Providence. After that junior year in Wyoming, the thought of being married to a lawyer in Providence gave me claustrophobia."

Their first child was a little white-haired kid named Frances, who arrived in Centralia, Illinois, where her father was based while he worked for Shell. Centralia was as rough as a frontier town, but "gone sour." Much of what happened there offended David's sense of fair play. In his words: "It was a boom town, full of run-out seed. There hadn't been a fair killing there since 1823." There was union trouble. The hod carriers were trying to organize the oil drillers. The drillers resisted, and had no intention of paying what they regarded as tribute. The hod carriers attacked. They scalped a driller with a hunting knife and then broke his bones with hammers. "That was macho stuff, to them," David comments. "They played rough. They were a mean bunch of bastards." In some of the towns he visited were signs that said "No Dogs or Oilmen." He posed as a travelling salesman.

In Tennessee, he was sometimes mistaken for a revenue agent, which could have led to an unpleasant fate. And on one occasion he was taken for a railroad detective by some fugitives from the law. Jane happened to be with him, and as they made their way along the tracks, pausing like detectives to examine the rock, the fugitives—who had dumped into a railroad cut a corpse that had needed killing—were watching from the woods. They drew beads with their rifles but held their fire. They didn't want to include Jane. Eventually, David learned all this from the fugitives themselves, and he asked them if they were not made uneasy by the discovery that they might have killed an innocent man. They gave him a jug of sorghum.

In the evening in Centralia, David could read his newspaper by the light of gas flares over the oil fields. The company was burning off the gas because, at the time, it lacked economic value. This impinged his Scottish temper. "I don't consider that good steward-ship," he explains. "We're stewards here—of land and resources. If you gut the irreplaceable resources, you're not doing your job. There were thousands of flares in Centralia. You could see them for a hundred miles." He was troubled as well by the secrecy of the oil com-pany, which was otherwise an agreeable employer. As a scientist, he believed in the open publication of re-search, and meanwhile his work was being locked in a safe for the benefit of one commercial interest. More-over, he moved around so much that in the first two years of his daughter's life she had been in thirteen states, while he was "looking for oil for some damn fool to burn up on the road." With Jane, he reached an obvious conclusion: "There has to be more to life than this."

He resigned to return to Wyoming with the U.S. Geological Survey, at first to pursue an assignment critical to the Second World War. The year was 1942, and the United States was desperately short of vana-dium, an alloy that enables steel to be effective as armor plate. Working out of Afton, in the Overthrust Belt, he looked for the metal in Permian rock. He first identified vanadium habitat (where it was in beds of black shale), and then—in winter—built a sawmill and cabins and made his own timbers for eight new mines. Afton was a Mormon redoubt. The municipal patriarch had thirty-four children. The Loves and the haberdasher Isidor Schuster were the only Gentiles in town. Love enlisted some Mormon farmers and taught them how to mine. They worked in narrow canyons, where avalanches occurred many times a day, coming two thousand feet down the canyon walls. David was caught in one, swept away as if he were in a cold tornado, and badly injured. He lay in a hospital many weeks. Not much of him was not damaged. His complete statement of the diagnosis is "It stretched me out."

After the mines were well established, the Loves moved to Laramie, where he set up the field office that has remained in Laramie more than forty years. His children grew up there. Frances now teaches French in public schools in Oklahoma. The Loves' two sons—Charles and David—are both geologists. Barbara teaches English at the University of Wyoming. Gradually —as regional field offices have been closed and geologists have been consolidated in large federal centers in Menlo Park (California), Reston (Virginia), and Denver—Love has become vestigial in the structure of the Survey. He has resisted these bureaucratic winds even when they have been stiffer than the winds that come over the Medicine Bows. "The tendency has been to have all the geologists play in one sand pile," he explains diplomatically. And here his friend Malcolm McKenna, curator of vertebrate paleontology at the American Museum of Natural History, takes up the theme: "Dave chose to stay where the geology is, and not to go up the ladder. He was so competent the Survey tried to get him to go out of Wyoming, but he wouldn't go. His is one of the few field offices left. The Survey gets information from people in addition to providing it. People stop in to see Dave. When he goes, the office in Laramie will close—and that will be a loss to Wyoming. While a whole bunch of people sit in little cubicles in Denver, Dave is close to the subject. He can walk out the door in the morning and do important stuff."

Love says that a part of his job is to find anything from oil to agates, and then, in effect, say, "Fly at it, folks," to the people of the United States. Within the law, he is free to resign and then fly at something himself, but—whether by oil, uranium, gemstones, or gold—time after time he has not so much as been tempted. Very evidently, he is not interested in money, and would not have joined the Survey in the first place if its services had been limited to commerce. The Survey evaluates the nation's terrain for academic purposes as well, there being no good way to comprehend any one aspect of geology without studying the wider matrix in which it rests. Within the geologic profession, the Survey has particular prestige—as much as, or even more than, the geology faculties of major universities, where chair professors have been known to mutter about the U.S.G.S., "They think they are God's helpers." Academic geologists tend to look upon the Survey as "stuffy." And, as Love discovered long ago, there is such an authoritarian atmosphere in the Survey—so much review of anything to be published, and so much hierarchical attention to a given piece of work—that sometimes when it is all done you cannot see the science for the initials that cover the paper. In its hundred and some years, the Survey has become so august that

McKenna refers to it as "an inertial organization, a
remnant of medieval scholasticism," but goes on to say,
"University people have two months a year; company
people are restricted. The Survey can do things no one
else can do." Many people in the profession tend to
think that a geologist who has not at some point worked
for the Survey has not been rigorously trained, for a
geologist seeking field experience can obtain it in such
quantity and variety nowhere else.

Love also established a base in Jackson Hole—a
small house, eventually a couple of cabins. This would
be the point of orientation for much of his summer field
work. His absorptions over the years would take him to
every sector of Wyoming, to other parts of the Rockies,
and elsewhere in the world. Always, though, from his
earliest days in geology, he would be drawn and drawn
back to the Teton landscape—to the completeness of
its history, the enigmas of its valley. To come to an
understanding of one such scene is to understand a
great deal about the geologic province of which it is a
part, and more than any other segment of the Rockies
he assigned himself to investigate the story of Jackson
Hole.

❧ "Hole" was a term used by the earliest whites to
describe any valley that was closely framed by very
high mountains. It was used by David Jackson, who

essentially had his valley to himself, running his trap lines in the eighteen-twenties in the afternoon shadows of the Teton Range. Over time, bands of outlaws followed him, then cattlemen, and eventually homesteading farmers, whose fences invaded the rangeland, creating incendiary tension and setting the scene for the arrival of Shane, who came into the valley wearing no gun and "riding a lone trail out of a closed and guarded past." A farmer offered him employment, and he accepted—earnest in his quest for a peaceful life. The farmer asked Shane almost nothing of his history but felt he could trust him and imagined a number of ways in which the man might be needed on the farm. Deep in the stranger's saddle roll was an ivory-plated Colt revolver that came out of its holster with no apparent friction, had a filed-down hammer and no front sight, and would balance firmly on one extended finger. The farmer's young son quite innocently discovered the gun one day, and hurried to his father.

"Father, do you know what Shane has rolled up in his blankets?"

"Probably a gun."

"But—how did you know? Have you seen it?"

"No. That's what he would have."

"Well, why doesn't he ever carry it? Do you suppose maybe it's because he doesn't know how to use it very well?"

"Son, I wouldn't be surprised if he could take that gun and shoot the buttons off your shirt with you a-wearing it and all you'd feel would be a breeze."

Shane, of course, was a fictional character, but the era he represented was a stratum of the region. In the opening words of the novel, by Jack Schaefer, "He rode into our valley in the summer of '89." He also glanced "over the valley to the mountains marching along the horizon." The geography is vague, but Schaefer evidently had in mind a place beside the Bighorn Mountains. When Hollywood took up the story, though, and prepared to spread it from Cheyenne to Bombay, the valley that Shane would ride into seemed an almost automatic choice. Its floor, as he slowly moved across it, was generally as flat as the bottom of a lake. Incongruous in its center were forested buttes, with clear cold streams running past them. In many places, the flatness was illusory, for there was random undulation and, for no apparent reason, a lyrical quilting of stands of dark pine and broad open stretches of pale-green sage. There were ponds, some of them warm enough to hold trumpeter swans for the winter; and lying against the higher mountains were considerable lakes. Mountains were everywhere. On three sides of the valley, they went up in fairly stiff gradients—the Mt. Leidy Highlands, the Gros Ventre Mountains, the Snake River Range. On the western side—without preamble, without foothills, with a sharp conjunctive line at the meeting of flat and sheer—were the Tetons, which seemed to have lifted themselves rapidly past timberline in kinetic penetration of the sky. The Tetons resemble breasts, as will any ice-sculpted horn—Weisshorn, Matterhorn, Zinalrothorn—at some phase in the progress of its making. Hollywood cannot resist the Tetons. If you have seen Western movies, you have seen the Tetons. They have appeared in the background of countless pictures, and must surely be the most tectonically active mountains on film, drifting about, as they will, from Canada to Mexico, and from Kansas nearly to the coast. After the wagon trains leave Independence and begin to move westward, the Tetons soon appear on the distant horizon, predicting the beauty, threat, and promise of the quested land. After the wagons have been moving for a month, the Tetons are still out there ahead. Another fortnight and the Tetons are a little closer. The Teton Range is forty miles long and less than ten across—a surface area inverse in proportion not only to its extraordinary ubiquity but also to its grandeur. The Tetons—with Jackson Hole beneath them—are in a category with Mt. McKinley, Monument Valley, and the Grand Canyon of the Colorado River as what conservation organizations and the Washington bureaucracy like to call a scenic climax.

In the Teton landscape are forms of motion that would not be apparent in a motion picture. Features of the valley are cryptic, paradoxical, and bizarre. In 1983, divers went down into Jenny Lake, at the base of the Grand Teton, and reported a pair of Engelmann spruce, rooted in the lake bottom, standing upright, enclosed in eighty feet of water. Spread Creek, emerging from the Mt. Leidy Highlands, is called Spread Creek because it has two mouths, which is about as common among creeks as it is among human beings. They are three miles apart. Another tributary stream is lower than the master river. Called Fish Creek, it steals along the mountain base. Meanwhile, at elevations as much as fifteen feet higher—and with flood-control levees to keep the water from spilling sideways—down the middle of the valley flows the Snake.

One year, with David Love, I made a field trip that included the Beartooth Mountains, the Yellowstone Plateau, the Hebgen earthquake zone of the Madison River, the Island Park Caldera, and parts of the Snake River Plain. Near the end of the journey, we came over Teton Pass and looked down into Jackson Hole. In a tone of sudden refreshment, he said, "Now, there is a place for a kid to cut his eyeteeth on dynamic geology."

Among others, he was referring to himself. He rode into the valley in the summer of '34. Aged twenty-one, he set up a base camp, and went off to work in the mountains. There were a number of small lakes among the Tetons at altitudes up to ten thousand five hundred feet—Cirque Lake, Mink Lake, Grizzly Bear Lake, Ice-floe Lake, Snowdrift Lake, Lake Solitude—and no one knew how deep they were or how much water they might contain. The Wyoming Geological Survey wanted to know, and had offered him a summer job and a collapsible boat. He climbed the Tetons, and rowed the lakes, like Thoreau sounding depths on Walden Pond. He likes to say that the first time he was ever seasick was above timberline. If the Teton peaks were like the Alps—a transplanted segment of the Pennine Alps—there was the huge difference that just up the road from the Pennine Alps there are no geyser basins, boiling springs, bubbling muds, or lavas that froze in human time. His base camp was on Signal Mountain—by Teton standards, a hill—rising from the valley floor a thousand feet above Jackson Lake. More than fifty summers later, one day on Signal Mountain he said, "When I was a pup, I used to come up here to get away from it all."

I said, "By yourself?"

And he answered, "Oh, yes. Always. No con-
cubines. I've always been pretty solitary. I still am."

Gouging around the mountains in his free time—
and traversing the valley—he would get off his horse
here and again, sit down, and think. ("You can't do
geology in a hurry.") On horseback or on foot—from
that summer forward, whenever he was there—he
gathered with his eyes and his hammer details of the
landscape. If he happened to come to a summit or an
overlook with a wide view, he would try to spend as
much of a day as possible there, gradually absorbing
the country, sensing the control from its concealed and
evident structure, wondering—as if it were a formal
composition—how it had been done. ("It doesn't matter
that I don't know what I'm looking at. Later on, it becomes clear—maybe. And maybe not. You try to put
the petals back on the flower.") Some of those summits
had not been visited before, but almost without exception he did not make a cairn or leave his name. ("I
left my name on two peaks. When you're young and
full of life, you do strange things.") Having no way to
know what would or would not yield insight, he noticed
almost anything. The mountain asters always faced east.
Boulders were far from the bedrock from which they
derived. There was no quartzite in any of the surrounding mountains, but the valley was deeply filled with
gold-bearing quartzite boulders. He discovered many
faults in the valley floor, and failed for years to discern
among them anything close to a logical sequence. There
were different episodes of volcanism in two adjacent
buttes. From high lookoffs he saw the barbed headwaters of streams that started flowing in one direction

and then looped about and went the other way—the sort of action that might be noticed by a person carrying water on a tray. Something must have tilted this tray. From Signal Mountain he looked down at the Snake River close below, locally sluggish and ponded, with elaborate meanders that had turned into oxbows— the classic appearance of an old river moving through low country. This was scarcely low country, and the Snake was anything but old. Several miles downstream, it took a sharp right, straightened itself out, picked up speed, and turned white. Looking down from Signal Mountain, he also noticed that moose, elk, and deer all drank from one spring just before their time of rut, crowding in, pushing and shoving to get at it ("They honk and holler and carry on"), ignoring the nearby waters of river, swamp, and lake. He named the place Aphrodisiac Spring. Over the decades, a stretch at a time, he completely circumambulated the skyline of Jackson Hole, camping where darkness came upon him, casting grasshoppers or Mormon crickets to catch his dinner. There were trout in the streams as big as Virginia hams. Sometimes he preferred grouse. ("I could throw a geology hammer through the air and easily knock off a blue grouse or a sage chicken. In season, of course. Hammer-throwing season. In the Absarokas, I threw at rattlesnakes, too. I don't kill rattlesnakes anymore. I've come to realize they're a part of the natural scene, and I don't want to upset it.") He carried no gun. He carries a bear bell instead. One day, when he forgot the bell, a sow grizzly stood up out of nowhere—six feet tall—and squinted at him. Suddenly, his skin felt dry and tight. ("Guess who went away.") A number of times, he was charged by moose. He climbed a tree. On one occasion, there was no tree. He and the moose were above timberline. He happened to be on the higher ground, so he rolled boulders at the moose. One of them shattered, and sprayed the moose with shrapnel. ("The moose thought it over, and left.")

The Gros Ventre River entered the valley almost opposite the high Teton peaks. A short way up the Gros Ventre was a denuded mountainside, where seventy-five million tons of rock had recently avalanched and dammed the river. He saw glacial grooves running north-south, and remembered the levees that kept the Snake from spilling west. This suggested to him that the valley floor had tilted westward since the glacier went by. Curvilinear pine-covered mounds cupped the valley's various lakes and held them close to the Tetons. Each lake was at the foot of a canyon. Evidently, alpine glaciers had come down the canyons to drop their moraines in the valley and, melting backward, fill the lakes. Some of the effects of ice were as fresh as that; others were less and less discernible, dating back from one episode of glaciation to another, separated by tens of thousands of years. Love's son Charlie, who now teaches geology and anthropology at Western Wyoming Community College, was hiking one day in 1967 along the ridgelines of the Gros Ventre Mountains when he discovered boulders whose source bedrock was fifty miles away in the Absaroka Range. If they were glacial, as they seemed to be, they recorded an episode until then unknown, and of greater magnitude than any other. The evidence remains scant, but what else could have carried those boulders fifty miles and set them down on mountain summits at ten thousand feet? David answered the question by coining the term "ghost glaciation."

From lookoffs in and around Jackson Hole, the view to the north concluded with a high and essentially level tree-covered terrain that seemed to be advancing from the direction of Yellowstone, as indeed it had done, spreading southward, concealing the earlier topography, filling every creek bed, pond, and gulch. When he first rode in that terrain, he saw with no surprise that the rock was rhyolite, which has the same chemistry as granite but not its crystalline texture, because rhyolite cools quickly as a result of pouring out upon the surface of the earth. This rhyolite, in a fiery cloud rolling down from Yellowstone, had buried the north end of the Teton Range, where it split and flowed along both sides.

From one end to the other of the valley were outcrops that from a distance looked like snow. Close up, they were white limestone, white shale, and white ash. After noting strikes and dips, and compiling the data, he calculated the thickness of the deposit as approximating six thousand feet. Top to bottom, it was full of freshwater clams and snails, and some beavers, aquatic mice, and other creatures that live in shallows. So the valley had been filled with a lake. The lake was always shallow. Yet its accumulated sediments were more than a mile thick. There was no rational explanation—unless the floor of the valley was steadily sinking throughout the life of the lake.

Volcanic rocks around the valley were white, brown, red, purple, and numerous hues of yellow and

green. Quartzite boulders—stream-rounded, and scattered far and wide—had come from a source far to the
northwest, in Idaho, and could not have been trans-
ported by ice. In the Mt. Leidy Highlands and along
the eastern edge of Jackson Hole he saw other boulders,
larger than human heads. Like the quartzites, they
asked questions that, for the time being, he could not
answer. He found black and gray sediments of the
Cretaceous seas. He measured them, and they were
two miles thick. Just above them in time, he found coal.
In red and salmon rock nearby were the small tracks
and tiny bones of dinosaurs. Larger ones, too. There
were beds of marine phosphate. He collected cherty
black shales, pure dolomites, dark dolomites, the massive sandstones of an ocean beach. He went into blue-
gray caves in beautiful marine limestone. He found
mud-crack-bearing shales. He saw mounds resembling
anthills, which had been built by blue-green algae.

Chipping with his hammer, he bagged folded and
fractured schist, amphibolite, and banded gneiss—and
granite that had come welling up as magma, intruding
these older rocks at a time when they were far below
the earth's surface, a time that was eventually determined by potassium-argon dating. The time was 2.5
billion years before the present. Therefore, the rock
that the granite invaded was a good deal older, but it
had been metamorphosed, and there was no telling how
long it had existed before it was changed—how far it
reached back toward the age of the oldest dated rock
on earth: 3.8 billion.

In these lithic archives—randomly assembled, subsequently arranged and filed—was a completeness in

[135]

every way proportionate to the valley's unexceedable beauty. From three thousand million years ago to the tectonically restless present, a very high percentage of the epochs in the history of the earth were represented. It was no wonder that a geologist would especially be drawn to this valley. As he moved from panorama to panorama and outcrop to outcrop—relating this rock type to that mountain, this formation to that river— David gradually began to form a tentative regional picture, and after thirty years or so had placed in sequential narrative the history of the valley. When new evidence and insight came along, what had once seemed logical sometimes fell into discard. When plate tectonics arrived, he embraced its revelations, or accommodated them, but by no means readily accepted them. He wrote more than twenty professional papers on subjects researched in the vicinity, and, with his colleague John Reed, published a summary volume for the general public called "Creation of the Teton Landscape." When the Department of the Interior honored him with a Citation for Meritorious Service, it said, in part, that he had "established the fundamental stratigraphic and structural framework for a region." In short, he had put the petals back on the flower.

And it was some flower. The Teton landscape contained not only the most complete geologic history in North America but also the most complex. ("One reason I've put in a part of my life here is that we have so much coming together. I don't want to waste my time. I can make more of a contribution by concentrating here than on any other place.") After half a century with the story assembling in his mind, he can roll it like a Roman

scroll. From the Precambrian beginnings, he can watch
the landscape change, see it move, grow, collapse, and
shuffle itself in an intricate, imbricate manner, not in
spatial chaos but by cause and effect through time. He
can see it in motion now, in several ways responsively
moving in the present—its appearance indebted to the
paradox that while the region generally appears to have
been rising the valley has collapsed.

Splitting the wall of the Tetons is a diabase dike a
hundred and fifty feet wide, running like a dark streak
of warpaint straight up the face of the mountains.
Diabase: a brother of gabbro, a distant relative of
granite. Four miles below the surface of the earth, the
space occupied by this now solid dike was once a fissure
through which the dark rock flowed upward as magma.
At the same point in the narrative—1.3 billion years
before the present, in the age of the Precambrian called
Helikian time—marine beaches are not far to the west,
and beyond them is a modest continental shelf. There is
no Oregon, no Washington, not much Idaho—instead,
blue ocean over ocean crust. Down toward the beaches
flow sluggish rivers across a featureless plain. Folded
and faulted schists and gneisses are bevelled under the
plain, preserving in their deformation compressive
crustal movements that have long since driven skyward
uncounted ranges that have worn away. The Helikian
beaches in their turn disappear, in burial becoming
sandstone, which in the heat and pressure of more
folding mountains is altered to quartzite. The mountains dissolve, and still another quiet plain vanishes
below waves. The water advances into this piece of the
world that will one day form as Jackson Hole. It lies

close to the latitude of Holocene Sri Lanka—or Malaya, or Panama—and is moving toward the equator. The water is warm but not always quiet or clear. Blue-green algae build mounds in the shallows. There is a drop in sea level. Polygonal mud cracks become ceramic in the tropical sun. The sea returns. The water is virtually transparent, and the skeletons of quintillions of creatures form a pure blue-gray limestone. Like Debussy's engulfed cathedral, the site comes up now and again into the light and air, but for the most part seas stay over it. Sands accumulate—broad, deep sands—but they preserve almost no fossil record, so not even David Love will ever say with certainty whether they are underwater or out in the air. (What he cannot say with certainty he will readily say without certainty, provided the difference is clear. He prefers not to be, as he likes to put it, "a man walking with one foot on each side of a fence." He thinks that some of those sands were terrestrial dunes and coastlines, reddened as oxides in the air.) Jackson Hole is close to the equator, and phosphates form in the shallow evaporating sea. Tidal flats appear—wide red flats, thickened by slow rivers coming from an uplift far to the east. In the muds are small tracks and tiny bones of dinosaurs. Rapidly—and possibly as a result of the breaking up of the earth's only continent—the region travels north, moving about a thousand miles in thirty million years. Big dunes form upon the flats: dry, windblown dunes—a Sahara in salmon and red, at the precise latitudes of the modern Sahara. The red sands in turn are covered by the Sundance Sea. Coming from the north, it not only buries the big dunes under mud and sand but covers them with galaxies of clams. When the water drops, floodplains emerge, and flooding rivers band the country— pink, purple, red, and green. Dinosaurs wander this chromatic landscape—a dinosaur as large as a corgi, a dinosaur as large as a bear, a dinosaur larger than a Trailways bus. Seas return, filled with a viciousness of life. Black and gray sediments pour into them from stratovolcanoes off to the west. In these times, the piece of sea bottom which is the future site of Jackson Hole overshoots the latitude of modern Wyoming and continues north to a kind of apogee near modern Saskatoon.

The land arches. Deep miles of sediments lying over schists and granites rise and bend. The seas drain eastward. The dinosaurs fade. Mountains rise northwest, rooted firmly to their Precambrian cores. Braided rivers descend from them, lugging quartzite boulders, and spreading fields of gold-bearing gravel tens of miles wide. Other mountains—as rootless sheets of whole terrain—appear in the west, sliding like floorboards, overlapping, stacking up, covering younger rock, colliding with the rooted mountains, while to the east more big ranges and huge downflexing basins appear in the random geometries of the Laramide Revolution.

For all that is going on around it, the amount of activity at the site of Jackson Hole is relatively low. Across the future valley runs a northwest-trending hump that might be the beginnings of a big range but is destined not to become one. Miles below, however, a great fault develops among the Precambrian granites, amphibolites, gneisses, and schists—and a crustal block moves upward at least two thousand feet, stopping, for the time being, far below the surface.

New volcanoes rise to the north and east. Fissures spread open. Materials ranging from viscous lavas to flying ash obliterate the existing topography. Streams disintegrate these materials and rearrange them in layers a few miles away. So far, these scenes—each one of which is preserved in the rock of Jackson Hole—have advanced to a point that is 99.8 per cent of the way through the history of the earth, yet nothing is in sight that even vaguely resembles the Tetons. The Precambrian rock remains buried under younger sediments. At the surface is a country of undramatic hills. The movements that brought the Overthrust Belt to western Wyoming—and caused the more easterly ranges to leap up out of the ground—have all been compressional: crust driven against crust, folded, faulted, and otherwise deformed. Now the crust extends, the earth stretches, the land pulls apart—and one result is a north-south-trending normal fault, fifty miles long. On the two sides of this fault, blocks of country swing on distant hinges like a facing pair of trapdoors—one rising, one sagging. The rising side is the rock of the nascent Tetons, carrying upward on its back the stratified deposits of half a billion years. One after another, erosion shucks them off. Even more rapidly, the east side falls—into a growing void. Magma, in motion below, is continually being drawn toward volcanoes, vents, and fissures to the north. Just as magma moving under Idaho is causing land to collapse and form the Snake River Plain, magma drawn north from this place is increasing the vacuity of Jackson Hole. As the magma reaches Yellowstone, it rises to the surface, spreads out in all directions, and in a fiery cloud rolls down from Yellowstone to bury the north end of the Tetons, where it splits and flows along both sides. The descending valley floor breaks into blocks, like ice cubes in a bucket of water. Some of them stick up as buttes. A lake now fills the valley—shallow, forty miles long— and in it forms a limestone so white it looks like snow. There are white shales as well, and water-laid strata of white volcanic ash. As these sediments thicken to a depth approaching six thousand feet, the lake that rests upon them is always shallow, and full of freshwater clams and snails, and some beavers and aquatic mice. While the lake is accepting sediment, the bottom of its bottom is sinking at the same rate. With a loud terminal hissing, lavas flowing down from Yellowstone cool in the lake as obsidian. Billows of hot sticky fog come down the valley as well. It cools as tuff. The big lake vanishes. In successive earthquakes, there is more valley faulting, damming the valley streams to form deep narrow lakes, which appear suddenly and as quickly go. Off the fast-rising block of mountains, erosion has by now removed fifteen thousand feet of layered sediments, and the Precambrian granites—with their attendant amphibolites, schists, and gneisses, and a vertical streak of diabase—are the highest rock below the sky. Bent upward against the flanks of the Precambrian are the broken-off strata of the Paleozoic era, and the broken-off strata of the Mesozoic era—serrated, ragged hogbacks, continually pushed aside. Perched on the

[*141*]

granite at the skyline is a bit of Cambrian sandstone that the weather has yet to take away. On the opposite side of the Teton Fault, the same sandstone lies beneath the valley. The vertical distance between the two sides of this once contiguous formation is thirty thousand feet.

That brings the chronicle essentially to the present, but still the blockish mountains look more like hips than breasts. Now off the Absarokas, off the Wind Rivers, off the central Yellowstone Plateau—and, to a lesser extent, down the canyons of the Tetons—comes a thousand cubic miles of ice. A coalesced glacier more than half a mile thick enters and plows the valley. The west side of this glacier scrapes along the Tetons above the level of the modern timberline. Melting away, the glacier leaves a barren ground of boulders. More ice comes—a lesser but not insignificant volume—and a third episode, which is smaller still. The ice cuts headward up canyons into the mountains, making cirques. As rings of cirques further erode, they form the spires known as horns. The ice signs the valley with lakes, and as it shrinks back into the mountains human beings have come to watch it go. Long after it is gone, the valley floor, continuing to be unstable as magmas are drawn north from below, drops even more. Big spruce go down with it—trees with diameters of five feet—and are enveloped by the water of Jenny Lake. The mountains jump upward at the same time, many feet in a few seconds near the end of the fourteenth century, emphasizing the fact that they are active in our time. In 1925, seventy-five million tons of rock fall into the Gros Ventre River. In 1983, the year that the trees are discovered at the bottom of Jenny Lake, an earthquake halfway up the Richter scale rumbles through Jackson Hole.

A geologic map is a textbook on one sheet of paper. In its cryptic manner—its codes of color and sign—it reflects (or should reflect) all the important research that has been done on any geologic topic within its boundaries. From broad formational measurements down to patterns in the fabric of the rock, a map should serve as an epitome of what is known and not known about a region, up to date. Regional maps have traditionally been presented state by state, and the dates they are up to vary: Nevada 1978, New York 1970, New Jersey 1910. On a geologic map, as on any scientific publication, the name of the person primarily responsible appears first. The job involves so many years and such a prodigious bibliography that the completion of a state geologic map can be regarded as the work of a lifetime, and David Love is only the second person in the history of American geology who has served as senior author of a state map twice (Wyoming 1955, Wyoming 1985). Geology is a descriptive, interpretive science, and conflict is commonplace among its practitioners. Where two or more geologists have come to divergent conclusions, Love has had to go out and rehearse their field work, in order to decide what to show on the map. People tend to become ornery when the validity of their assertions is challenged, and figuratively some of his colleagues have reached for their holsters, which may have been a mistake, as the buttons fell off their shirts and they felt a little breeze.

The 1955 Wyoming map set a standard for state geologic maps in the detail of its coverage, in its fossil-dating, in its delivery of the essence of the region—a standard set anew in the 1985 edition. In the words of Malcolm McKenna, of the American Museum of Natural History, "Most maps are patched together from various papers and reports. Dave has looked at all the rock. It's all in one mind. Most geologic maps are maps of time, not rocks. They will say something like 'undifferentiated Jurassic' and omit saying what the rocks are. There is little of that on Dave's map. Mapping is below the salt now. Yet you can't look at satellite photos for everything. You've got to have high-resolution basic mapping. You have to keep your hand in with the real stuff. When the solid foundations aren't there, geologists are talking complete mush. Dave is making sure the foundation is there. He does not write about geology from a distance. He does not sit in high councils figuring out how the earth works. He is field-oriented. Some geologists think field work is wheeling their machines out into the yard. Dave has his hand on the pulse. He knows geology from having found it out himself. He has set an example of the way geology is done—one hell of an example. To compete with Dave, you'd have to do a lot of walking."

Love once picked up a mail-order catalogue and saw an item described as "Thousand Mile Socks." He sent for them skeptically but later discovered that there was truth in the catalogue's claim. They were indeed thousand-mile socks. He had rapidly worn them out, but that was beside the point.

Years ago, almost anybody going into geology could look forward to walking some tens of thousands of miles and seriously studying a comparable number of outcrops. Geology, by definition, was something you did in the field. You sifted fine dirt for fossils the eye could barely see. You chiselled into lithified mud to remove the legs of dinosaurs. You established time-stratigraphic relationships as you moved from rock to rock. You developed a sense of structure from, among other things, your own mapping of strikes and dips. In the vernacular of geology, your nose was on the outcrop. Through experience with structure, you reached for the implied tectonics. Gradually, as you gathered a piece here, a piece there, the pieces framed a story. Feeling a segment of the earth, you were touching a body so great in its dimensions that you were something less than humble if you did not look upon your conclusions as tentative. Like many geologists, Love became fond of the Hindu fable of the blind men and the elephant (six blind men feel different parts of an elephant and come to six very different opinions of what it is), because the poem in a few short verses allegorized for him the history and the practice of his science. "We are blind men feeling the elephant," he would say, almost ritually, as a way of reminding anyone that the crust is so extensive and complicated—and contains so little evidence of most events in earth history—that every relevant outcrop must be experienced before a regional outline can so much as be suggested, let alone a global picture.

In recent years, the number of ways to feel the elephant has importantly increased. While the science has assimilated such instruments as the scanning transmission electron microscope, the inductively coupled plasma spectrophotometer, and the $^{39}Ar/^{40}Ar$ laser microprobe—not to mention devices like Vibroseis that thump the earth to reflect deep structures through data reported by seismic waves—the percentage of geologists has steadily diminished who go out in the summer and deal with rock, and the number of people has commensurately risen who work the year around in fluorescent light with their noses on printouts. This is the age of the analog geologist, who, like a watch with a pair of hands, now requires a defining word. For David Love, the defining word is "field." Whereas all geologists were once like him, they are no longer, and his division of the science is field geology. He is the quintessential field geologist—the person with the rock hammer and the Brunton compass to whom weather is just one more garment to wear with his thousand-mile socks, the geologist who carries his two-hundred-megabyte hard disk between his ears. There are young people following in his steps, people who still go out to scuff their boots and fray their jeans, but they have become greatly outnumbered by their contemporaries who feed facts and fragments of the earth into laboratory machines—activity that field people describe as black-box geology. Inevitably, some touches of tension have appeared between these worlds:

"Who is the new structural geologist?"

"Dorkney."

"Is he a field-oriented person?"

"He's a geophysicist, but he's a good guy."

"That would be difficult."

Black-box geologists—also referred to as office geologists and laboratory geologists—have been known to say that field work is an escape mechanism by which their colleagues avoid serious scholarship. Their remarks may rarely be that overt, but the continuing relevance of field geology is not—to say the least—universally acknowledged. Some laboratory geologists, on the other hand, are nothing less than eloquent in expressing their symbiosis with people of wide experience out in the terrain. "I spend most of my time working on computers and waving my arms," a geophysicist once said to me, adding that he required the help of someone's field knowledge as a check, and without it would be in difficulty. "Without such people there would be no such thing as a geological enterprise," he went on. "Every box of samples that comes into the lab should include a worn-out pair of field boots. There's a group of senior geologists who have met on the outcrops and share a large body of knowledge. They paste together different perceptions of the world by visiting each other's areas. When I meet them, I chat them up like the guys at the corner store, because what I do is conceptual and idealized, and I'd like to know that it relates to what they have seen. These people are generally above fifty. Their kind is being diminished, which is a major intellectual crime. It has to do with the nature of science and what we're doing. Reality is not something you capture on a blackboard."

Such sentiments notwithstanding, within university geology departments black-box people tend to outvote field people on questions of curriculum and directions of research, and to outperform them in pursuit of funds. "The black-box era has been caused by the availability of money for esoteric types of work," Love remarked one day. "The Department of Defense, the National Science Foundation, and so forth have had money to spend on—let's say—unusual quests. The experience you get from collecting rocks in the field is lost to the lab geologist. For example, there's a boom in remote-sensing techniques—in satellite imagery. From that, you get a megapicture without going into the field. But it's two-dimensional. To get the third dimension—to study what's underground—they consult another sacred cow, which is geophysics. They can make a lot of these interpretations in the office. They can go off the mark easily, because for field relationships they often rely on data collected years ago. They use samples from museums, or samples collected by somebody else—perhaps out of context. I'm afraid I'm rather harsh about it, but we see misinterpretations, because of lack of knowledge of field relationships. Many of the megathinkers are doing their interpretations on the basis of second- and third-hand information. The name of the game now is 'modelling.' A lot of it I can't see for sour owl shit. How can you write or talk authoritatively about something if you haven't seen it? It isn't adequate to trust that the other guy is correct. You should be able to evaluate things in your own right. Laboratory geology is where the money is, though. The money is in the black box. I think eventually it will get out. You can't blame the kids for doing this kind of office research when they're financed. I don't want to do it myself. Putting the geologic scene into a broad perspective is for me more satisfying. I want to know what's over the next hill."

He was saying some of this in the Mt. Leidy Highlands one day when we were sitting on an outcrop at ninety-two hundred feet and looking at a two-hundred-and-seventy-degree view that ran across the pinnacled Absarokas to a mountain of lava of Pleistocene age and then on up the ridgeline of the Continental Divide to the glaciers and summits of the Wind River Range, thirty-eight feet higher than the Tetons. The skyline sloped gently thereafter, flattened, and became the sub-summit surface of Miocene age, the level of maximum burial. There followed, across the southern horizon, the whole breadth of the Gros Ventre Mountains, with afternoon light on bright salmon cliffs of Nugget sandstone, at least four hundred feet high. The eye moved west over other summits and ultimately came to rest on the full front of the Tetons. We looked at it all for a considerable time in silence. Love said he liked this place because he could see so much from it, and had stopped here many times across the decades, to lean against a piñon pine and sort through the country, like an astronomer with the whole sky above him sorting through the stars. He also said, reflectively, "I guess I've been on every summit I can see from here."

Below us was Dry Cottonwood Creek. It ran southeast several miles, and then turned through a tight bend to head west toward the Tetons. We could see other streams almost identical in configuration, like a collection of shepherd's crooks. "The land tilted east, and then south, before it tilted west," Love said. "This is the tilting block that stops at the foot of the Tetons. The barbed streams are evidence that the hinge is east of us here. The hinge is probably the Continental Divide. We can learn a lot from streams. They're so sensitive. They respond to the slightest amount of tilting. I think this is underestimated." Pointing down to some sandstone ledges along the bank of Dry Cottonwood Creek, he said that Indians had frequently camped there because long ago the stream was so full of trout you could reach in under the ledges and catch them with your hand. He asked if I knew why the water was so clear. "There's no shale upstream," he said. "No fines to contaminate it. If you look at a stream, you can see in the sediments the whole history of a watershed. It's as plain as the lines on the palm of your hand."

On the way up to the lookoff, we had stopped at a spring, where I buried my face in watercress and simultaneously drank and ate. Love said that F. V. Hayden, the first reconnaissance geologist in Wyoming Territory, also happened to be a medical doctor, and he went around dropping watercress in springs and streams to prevent scurvy from becoming the manifest destiny of emigrants. Hayden, who taught at the University of Pennsylvania, led one of the several groups that in 1879 combined to become the United States Geological Survey. When he came into the country in the late eighteen-fifties, he was so galvanized by seeing the composition of the earth in clear unvegetated view that he regularly went off on his own, moved hurriedly from outcrop to outcrop, and filled canvas bags with samples. This puzzled the Sioux. Wondering what he could be collecting, they watched him, discussed him, and finally attacked him. Seizing his canvas bags, they shook out the contents. Rocks fell on the ground. In that instant, Professor Hayden was accorded the special status that all benevolent people reserve for the mentally disadvantaged. In their own words, the Sioux named him He Who Picks Up Rocks Running, and to all hostilities thereafter Hayden remained immune.

I remarked at the spring that Love was having nothing to drink.

He said, "If I drink, I'll be thirsty all afternoon."

And now, on the high outcrop, turning again from the Eocene volcanic Absarokas to the Wind River Range (the supreme expression in Wyoming of the Laramide Orogeny) and on to the newly risen Tetons (by far the youngest range in the Rockies), I mentioned the belief of some geologists that of all places in the world the Rocky Mountains will be the last to be deciphered in terms of the theory of plate tectonics.

"I don't think I would necessarily agree," Love said. "I think it is one of the more difficult ones, yes. I've thought a lot about it. At this stage, I'm uncomfortable with a direct tie-in. Until we have a detailed chronology of all the mountains, how can we plug them into a megapicture of plate tectonics? I don't want to give a premature birth to anything."

Plate-tectonic theorists pondering the Rockies have been more than a little inconvenienced by the great distances that separate the mountains from the nearest plate boundaries, where mountains theoretically are built. The question to which all other questions lead is, What could have hit the continent with force enough to drive the overthrust and cause the foreland mountains to rise? In the absence of a colliding continent—playing the role that Europe and Africa are said to have played in the making of the Appalachians—theorists have lately turned to the concept of exotic terrains: island arcs like Japan slamming up against the North American mainland one after another, accreting what are now the far western states, and erecting in the course of these collisions the evidential mountains. Whatever the truth in that may be, a tectonic coincidence very much worth noting is that the development of the western mountain ranges begins at the same time as the opening of the Atlantic Ocean. In the middle Mesozoic, as the Atlantic opens, the North American lithosphere, like a great rug, begins to slide west, abutting, for the most part, the Pacific Plate. A rug sliding across a room will crumple up against the far wall.

"We're about a thousand miles from the nearest plate boundary," Love was saying. "We should not tie in the landscape here with events that have taken place along the coast. This doesn't neutralize or dispose of the theory of plate tectonics, but applied here it's incongruous—it's kind of like a rabbit screwing a horse. There is no evidence of plates grinding against each other here. The thrust sheets are probably symptoms of plate-tectonic activity fifty million years ago, but the chief problem is that tectonism is not adequately placed in a time framework here. Almost everybody now agrees that there is tremendous significance to plate tectonics —also that the concept is valid. Most people don't argue about that anymore. Our arguments come in the details. We should dissect all these mountain ranges before we get diarrhea of the pen trying to clue them in to plate theory. There's nothing wrong with ideas, with working hypotheses, but unsubstantiated glittering generalities are a waste of time. Most of the mega-thinkers are basing sweeping interpretations on pretty inadequate data. There are swarms of papers being written by people who have been looking at state and federal and worldwide geologic maps and coming to sweeping conclusions on how mountains were formed and what the forces involved were. Until we know the anatomy of each mountain range, how are we going to say what came up when—or if they all came up in one great spasm? You can't assume they're all the same. In order to know the anatomy of each mountain range, you have to know details of sedimentary history. To know the details of sedimentary history, you have to know stratigraphy. I didn't know until recently that stratigraphy is dead. Many schools don't teach it anymore. To me, that's writing the story without knowing the alphabet. The geologic literature is a graveyard of skeletons who worked the structure of mountain ranges without knowing the stratigraphy. In Jackson Hole in the late Miocene, you had a lake that collected six thousand feet of sediment, half of which was limestone that was chemically precipitated. There had to be a source. It came from broad exposures of Madison limestone in the ancestral Teton–Gros Ventre uplift, chemically dissolved and then precipitated with cool-climate fossils. Therefore, that lake lay under a cool, humid climate. First, a basin had to be created in which the material was deposited—a basin ultimately thirteen thousand feet deep to accommodate all the lake and river sediments we find there, which puts it two miles below sea level at a time when the region is supposedly uplifting. All this is basic to structure, and the structure is basic to tectonics. The Owl Creek Mountains and the Uinta Mountains trend east-west. Why? Why are their axes ninety degrees from what you would expect if the tectonic force came from the west? You can do a torsion experiment with a rubber sheet and get folds in various directions—you can get east-west uplifts in the rubber sheet—but I would not say that is conclusive. You have mountains foundering. You have thrusting in the Laramide and sinking forty to fifty million years later, causing parts of basins to tilt this way and that like broken pieces of piecrust. The Granite Mountains were once as high as the Wind Rivers. Why did they go down? How did they go down? I don't think we're ready yet to put together a real megapicture. The 1985 geologic map of Wyoming consists of eighty-five-per-cent new mapping since 1955. The amount remapped shows how much new information was acquired in thirty years. The Big Picture is not static. It will always include new ideas, new tectonics, new stratigraphy. This information is an essential part of the megathinking of the plate-tectonics people, and twenty-five, fifty, a hundred years from now it will be very different."

West of Rawlins on Interstate 80, Love and I in the Bronco came into a region a good deal flatter than most of Iowa, with so little relief that there were no roadcuts for more than fifty miles. Among dry lakebeds dimpling the Separation Flats, our altitude was seven thousand feet, yet the distant horizon was close to the curve of the earth. In this unroughened milieu, we passed a sign informing us that we were crossing the Continental Divide. So level is the land there that the divide is somewhat moot. Cartographers seem to have difficulty determining where it is. Its location will vary from map to map. Moreover, it frays, separates, and, like an eye in old rope, surrounds a couple of million acres that do not drain either to the Atlantic or the Pacific—adding ambiguity to the word "divide."

With respect to underlying strata, we were running along the crest of an arch between two sedimentary basins, although nothing on the surface suggested that this was so, for the basins were completely filled. The flats to our left were the Washakie Basin, to our right the Great Divide Basin—each like a bowl brimming over with Eocene alluvial soup. Younger deposits —maybe a mile's thickness—had long since been washed or blown away, leaving a fifty-million-year-old surface on which anything modern might fall.

A shepherd on horseback stood out against the sky, more so than his sheep. Even from a distance, he looked cold and uncomfortable. On this robust May afternoon, gray clouds, moving fast, were beginning to throw down hail. Love turned off the interstate, and the vehicle bucked south for a couple of miles in drab brown ruts

[155]

that suddenly turned bright, almost white, as the ground jumped forward in time roughly fifty million years. This patch of thirty or forty acres was all that remained locally of a volcanic-ash fall that had covered large parts of Wyoming, Colorado, Kansas, and Nebraska and had reached as far as Texas. Potassium-argon dating has established the event at six hundred thousand years before the present—far along in Pleistocene time, and an extremely recent date in the history of the world when you reflect that the age of the earth is eight thousand times the age of that ash fall, just as the United States of America is eight thousand times as old as something that happened in the middle of last week. The ash—consisting of very small shards of glass—had travelled about two hundred miles downwind from its volcanic source. Two hundred miles downwind from Mt. St. Helens, in the state of Washington, the amount of ash that has accumulated as a result of Mt. St. Helens' recent eruptions is three inches. The ash here at the Continental Divide was sixty feet thick. A hundred and more miles northwest are remnants of the same fallout, suggesting the dimensions of the great regional blanket of six hundred thousand years ago, now almost wholly lost to erosion. Love said cryptically, "We have to assume it fell on saint and sinner alike." It had not been milled around by streams. It was a pure ash, distinctly windborne, containing no sand, no clay. He said that some woolly-mammoth bones had been found not far away, and with them as a minor exception this ash marked the only firm Pleistocene date in an area of twenty thousand square miles. After settling, it had not consolidated—as volcanic ash sometimes will, forming welded tuff. (The

Vesuvian air-fall ash that settled on Pompeii also flew too high to weld. Rising rapidly like smoke, it actually pooled up against the stratosphere. Pliny said it looked like a flat-topped Italian pine. The geological term for such an event is Plinian eruption.) A couple of hundred miles northwest of us were the paint pots and fumaroles, the geysers and calderas of Yellowstone. Love said that this Lava Creek ash represented one of the great out-pourings in Yellowstone history. The hail now was pelting us. It collected like roe on the brim of his Stetson. Love seemed to regard it as a form of light rain, as something that would not last even for six hundred thousand nanoseconds and was therefore beneath notice.

Most volcanoes and related phenomena—most manifestations of the sort represented by the surface history of Yellowstone—are lined up along boundaries of the twenty-odd plates that collectively compose the earth's outer shell. The plates, which are something like a sixtieth of the earth's radius, slide around on a layer of the mantle hot enough to be lubricious. Where plates spread apart (the Red Sea, the mid-Atlantic), fresh magma wells up to fill the gap. Where plates slide by one another (San Francisco, Jericho), the ground is torn and walls collapse. Where plates collide (Denali, Aconcagua, Kanchenjunga), impressive mountains form. In collision, one plate usually slides beneath the other, plunging—in the so-called subduction zone—as much as seven hundred miles. The material carried down there tends to melt, and to rise as magma, reaching the surface in volcanic form, as in the Cascade Range, the Andes, the Aleutians, and Japan. Yellowstone, with all its magmatic products and bubbling sulphurs, attracts special attention in the light of this story, because Yellowstone is eight hundred miles from the nearest plate boundary.

When the theory of plate tectonics coalesced in the nineteen-sixties, it erased numerous enigmas of long pedigree in the science, and drew together in a single narrative aspects of geology that almost no one had guessed were related. In their excitement, geologists found themselves revelling in the novelty that the whole earth suddenly made sense. Plate tectonics seemed to account for everything from earthquakes to volcanic eruptions, from the great deeps of the oceans to the great altitudes of mountains. Like any practical theory, though, it asked as well as answered questions—and not a few of the questions were inconvenient to the theory. Many had to do with volcanism. For example, why was the island of Hawaii pouring out lava in the dead center of the Pacific Plate? Similarly, if volcanoes were the products of subduction zones, where was the nearest subduction zone to the Tibesti Mountains of Saharan Chad? The Tibesti massif—a couple of thousand kilometres from the leading edge of the African Plate—consists of shield volcanoes like Mauna Kea and Mauna Loa. Where was the closest subduction zone to the chain of peaks that culminates in Mt. Cameroon, a stratovolcano fifteen hundred miles from the nearest plate boundary of any kind? Moreover, some of the fine old conundrums of geology—problems that antedated the plate-tectonics revolution—remained standing in its aftermath. What could explain the Canadian Shield? The South American Shield? The South African Shield? How could so much Precambrian rock lie close to sea level and not have been buried in a thousand million years? What, in recent time, had lifted the platform of the Rockies, causing their exhumation? Why were Love and I, there on the platform, not at sea level? What had lifted the Colorado Plateau, subjecting it to incision by canyon-cutting rivers? What explained flood basalts? Plate tectonics seemed to have no relevance to them. With plate theory, you would think you could predict the sedimentary history of continents, but you couldn't. Why were continental basins—the Michigan Basin, the Illinois Basin, the Williston Basin—several kilometres deep? If you expect a shieldlike situation as the ultimate scene, what could explain these anomalous deep basins? Oil people wanted to know most of all. They asked plate theorists, "What does plate tectonics tell us about these basins?" The answer was "Nothing." Why were the granites of New Hampshire relatively young, and therefore anachronistic in the Appalachian story? What explained great crustal swells, like random blisters on the ocean floor, rising high above the abyssal plains? What could explain Bermuda—a mountain summit seventeen thousand six hundred and fifty-nine feet above the Hatteras Abyssal Plain? What created the Marshall Islands, the Gilbert Islands, the Line Islands, the Tuamotu Archipelago—where corals veneer the peaks of twenty-thousand-foot mountains that tend to run in chains? Like Yellowstone, like Bermuda, like Hawaii, like Mt. Cameroon, they lie great distances from the nearest intersections of plates.

Yellowstone draws its name from rich golden splashes of chemically altered volcanic rock. The place smokes and spits—the effects of proximate magma. On a geologic map of North America, Yellowstone appears at the eastern end of a bright streak of volcanic debris, coming off it like a contrail, extending across Idaho. With distance from Yellowstone, rock on that track is progressively older, descending age by age to the Columbia River flood basalts, which emerged from the ground like melted iron in early Miocene time, spread out across three hundred thousand square kilometres (in some places two and three miles deep), filled the Columbia Valley, and pooled against the North Cascades. By comparing the dates of the rock, one could be led to conclude that the geologic phenomenon now called Yellowstone has somehow been moving east at a rate of two and a half centimetres a year. As it happens, that is the rate at which, according to plate-tectonic theorists, North America is moving in exactly the opposite direction. In increasing numbers, geologists have come to believe that in a deep geophysical sense Yellowstone is not what is moving. They believe that the great heat that has expressed itself in so many ways on the topographic surface of the modern park derives from a source in the mantle far below the hull of North America. They believe that as North America slides over this fixed locus of thermal energy the rising heat is so intense that it penetrates the plate.

The geologic term for such a place is "hot spot." The earth seems to have about forty of them—most older, and many less productive, than Yellowstone. Despite its position under thick continental crust, the Yellowstone hot spot has driven to the surface an amount of magma that is about equal to the over-all production of Hawaii, which has written a clear signature on the Pacific floor. Hawaii is the world's most preserved and trackable hot spot. You can see its geologic history on an ordinary map if the map shows even the rudiments of what lies below the sea. The Pacific Plate is moving northwest. It dives into the Japan Trench, the Aleutian Trench, and regurgitates the volcanic islands that lie on the far side. The plate used to move in a direction closer to true north, but forty-two million years ago it shifted course. Any hot spot now active under the Pacific Plate will produce islands or other crustal effects that appear to be moving in the opposite direction—southeast. Mauna Kea and Mauna Loa—the shield volcanoes that from seafloor to summit are the highest mountains on earth—stand close to the southeasterly tip of the Hawaiian Islands. The extremely eruptive Kilauea is making the tip. The islands become lower, quieter, older —the farther they lie northwest. Islands older still— defeated by erosion—now stand below the waves. These engulfed ancestors of Hawaii form a clear track in the Pacific crust for more than five thousand miles. When their age reaches forty-two million years, their direction bends north. Above the bend, they are known as the Emperor Seamounts. Ever older, they continue to the juncture of the Kuril and Aleutian Trenches, into which they disappear. The oldest of the Emperor Sea-mounts is Cretaceous in age. Mauna Loa, of course, is modern. Under the ocean forty miles southeast of Mauna Loa is Loihi, a mountain of new basalt, which has already risen about twelve thousand feet and should make it to the surface in Holocene time.

The ages of the Emperor Seamounts and the familial Hawaiian islands create the illusion that Hawaii is propagating southeast at a rate of nine centimetres a year, while the message from plate tectonics is, of course, that the Pacific Plate is what is moving. The speeds and directions of the plates have been established by a number of corroborant observations. Offsets in faults like the San Andreas have been measured as expressions of time. Places in California that were once side by side and are now four hundred miles apart are also separated by eleven million years. A great deal of ocean-crustal rock has been dredged up and radioactively dated. The ages have been divided by distance from the spreading center to determine the rate at which the rock has moved. Methods are being refined for annual measurements of plate motions by satellite triangulation. Hot spots provide one more way of calculating plate velocities, for hot spots are to the drift of plates as stars to navigation.

Conversely, it is possible to use established tectonic velocities to chart the tracks of hot spots with respect to the overriding plates. Given just one position and one date (the present will do), it is possible to say where, under the world, a hot spot would have been at any time across a dozen epochs. W. Jason Morgan, a geophysicist at Princeton, has sketched out many such tracks and reported them in various publications. Morgan can fairly be described as an office geologist who spends his working year indoors, and he is a figure of first importance in the history of the science. In 1968, at the age of thirty-two, he published one of the last of the primal papers that, taken together, constituted the plate-tectonics revolution. Morgan had been trained as a physicist, and his Ph.D. thesis was an application of celestial mechanics in a search for fluctuations in the gravitational constant. Only as a postdoctoral fellow was he drawn into geology, and assigned to deal with data on gravity anomalies in the Puerto Rico Trench. Fortuitously, he was assigned as well an office that he shared for two years with Fred Vine, a young English geologist who, with his Cambridge colleague Drummond Matthews, had discovered the bilateral symmetry of the spreading ocean floor. This insight was fundamental to the revolutionary theory then developing, and sharing that office with Fred Vine drew Morgan into the subject—as he puts it—"with a bang." A paper written by H. W. Menard caused him to begin musing on his own about great faults and fracture zones, and how they might relate to theorems on the geometry of spheres. No one had any idea how the world's great faults—like, say, the San Andreas and Queen Charlotte Faults—might relate to one another in a system, let alone how the system might figure in a much larger story. Morgan looked up the work of field geologists to learn the orientations of great faults, and found remarkable consistencies across thousands of miles. He tested them—and ocean rises and trenches as well—against the laws of geometry for the motions of rigid segments of a sphere. At the 1967 meeting of the American Geophysical Union, he was scheduled to deliver a paper on the Puerto Rico Trench. When the day came, he got up and said he was not going to deal with that topic. Instead, reading a paper he called "Rises, Trenches, Great Faults, and Crustal Blocks," he revealed to the geological profession the existence of plate tectonics. What he was saying was compressed in his title. He was saying that the plates are rigid—that they do not internally deform—and he was identifying rises, trenches, and great faults as the three kinds of plate boundaries. Subsequently, he worked out plate motions: the variations of direction and speed that have resulted in exceptional scenery. I once heard a friend of Morgan's ask him what he thought he could possibly do, if anything, as an encore to all that. Morgan is shy, and speaks softly, in accents that faintly echo a youth in Savannah, Georgia. "I don't know," he answered, with a shrug and a smile. "Prove it wrong, I guess."

Instead, he developed an interest in hot spots and the thermal plumes that are thought to connect their obscure roots in the mantle with their surface manifestations—a theory that would harvest many of the questions raised or bypassed by plate tectonics, and similarly collect in one story numerous disparate phenomena.

In 1937, an oceanographic vessel called Great Meteor, using a newly invented depth finder, discovered under the North Atlantic a massif that stood seventeen thousand feet above the neighboring abyssal plains. It was fifteen hundred miles west of Casablanca. No one in those days could begin to guess at the origins of such a thing. They could only describe it, and name it Great Meteor Seamount. Today, Jason Morgan, with other hot-spot theorists, is prepared not only to suggest its general origin but to indicate what part of the world has lain above it at any point in time across two hundred million years. Roughly that long ago, they place Great Meteor under the district of Keewaytin, in the Northwest Territories of Canada, about halfway between Port Radium and Repulse Bay. That the present

Great Meteor Seamount was created by a hot spot seems evident from the size and configuration of its base, which is a thousand kilometres wide and closely matches the domal base of Hawaii and numerous other hot spots. If a submarine swell is of that size, there is not much else it can be. That it was once, theoretically, somewhere between Port Radium and Repulse Bay is a matter of tracing and dating small circles on the sphere traversed by moving plates.

Keewaytin is in the center of the Canadian Shield. If the shield once had younger sediments on it, a hot spot underneath it would have lifted it up and cleaned it off, creating the enigma of the Canadian Shield. Morgan believes that various hot spots positioned in various eras under shield rock are what have kept it generally free of latter-day deposits. Stubborn fragments of the Paleozoic here and there on the shield suggest that this is so, as does the relatively modest number of meteorite craters. If the shield rock had been sitting there uncovered since Precambrian time, its surface could be expected to be more widely pockmarked, not unlike the plains of the moon.

Later in the Jurassic, the Great Meteor Hot Spot was under the west side of Hudson Bay, and in the early Cretaceous under Moose Factory, Ontario. All this is postulated not on any direct field evidence but simply on a charted extrapolation from an ocean dome nearly four thousand miles away. As time comes forward, however, the calculations place the hot spot—with its huge volumes of magma—under New Hampshire a hundred and twenty million years ago. The radioactively derived age of a good deal of granite in the White Mountains, so puzzlingly "anachronistic" in Appalachian history, is a hundred and twenty million years.

East of the North American continental shelf, lined up like bell jars on the Sohm Abyssal Plain, are the New England Seamounts. Their average height is eleven thousand feet. They are very well dated, and their ages decrease with distance east. Their positions and their ages—ninety-five million years, ninety million years, eighty-five million years—coincide with Morgan's mathematical biography of the Great Meteor Hot Spot.

A development that has greatly improved the precision of these measurements is argon-argon dating. A stream of neutrons in a nuclear reactor bombards a rock sample and causes a known fraction of its atoms of potassium to change into argon-39. Also in the sample are atoms of the isotope argon-40, which are unaffected by the bombardment and are the result of the natural decay of potassium through geologic time. The rate of decay is known and constant. The higher the proportion of argon-40, the older the rock. A mass spectrometer measures these ratios to establish a date. The older procedure known as potassium-argon dating—hitherto the best way of determining the age of something more than a few tens of thousands of years old—is done in two steps, requiring two samples. First, a chemical process determines how much potassium is present. Then a mass spectrometer looks at the second sample to see how much potassium has altered radioactively to become its daughter argon. The procedure suffers from the effects of weathering, which occur not only on the surface of rock but from grain to grain within. Argon slips away from weathered material, thus changing its over-all ratio to potassium and making any date determined by this method all the more approximate. Argon-argon dating is accomplished in the microscopic core of a single grain, beyond even the faintest disturbances of weather. The newer method is significantly more consistent and accurate than the older one. Results have shown—notably among the New England Seamounts—that where many potassium-argon dates fall into general approximation with Morgan's calculations, the dates derived by argon-argon follow the track exactly.

Eighty million years ago, in the Campanian age of late Cretaceous time, Great Meteor would have underlain the American-African plate boundary, the Mid-Atlantic Ridge. Since then, Great Meteor has cut a gentle curve southward through the African Plate. From late Cretaceous, Paleocene, and Eocene time, the path is as well defined as it is on the American side. After the Eocene, the hot spot made the big seamount that bears its name. Then it began to go cold, to evanesce, to fade like a shooting star.

Shooting star. Almost everyone who describes hot spots is tempted to reverse reality and go for illusion at the expense of fact—that is, to narrate the apparent travels of hot spots as if they were in motion leaving trails like shooting stars, instead of telling the actual story of slow crustal drift over the fixed positions of thermal plumes. Myself included. With words, it is much easier to move a hot spot than it is to move a continent. Here, for example, is the story of another of the world's hot spots told in terms of its illusory motion. With the flood basalts of Serra Geral, in southern Brazil, a hot spot is said to have begun in late Jurassic time. It moved east under Brazil for several million years and then crossed over to Africa, which at that time was not much separated from South America. It lifted mountains in Angola, and then, doubling back, headed southwest under the ocean to form the Walvis Ridge, a line of seamounts whose ages agree exactly with plate calculations for the hot spot, which stands in the ocean at present as Tristan da Cunha.

From the Serra Geral to the present island, the Tristan da Cunha hot-spot track is so well defined and dated that, as Morgan says, "it really ties down Africa." Not to mention South America.

An automatic inference from the theory is that hot spots perforating the same plates at the same times must make parallel tracks. On the floor of the Pacific, the tracks of the Line Islands, the Tuamotu Archipelago, the Marshalls, and the Gilberts parallel the track of Hawaii and the Emperor Seamounts. In the Atlantic, the Canary Islands have traced a curve parallel to Madeira's. Both are hot spots, and have left tracks that conform to Great Meteor. The Cape Verde Islands are a hot spot. A hundred and seventy million years ago, it was under New Hampshire, on a track nearly coincident with the later track of Great Meteor. The most voluminous intrusions of granite in the White Mountains are dated around a hundred and seventy million years. When Charles Darwin, on the Beagle, put in at the Cape Verde Islands, he had with him Charles Lyell's recently published "Principles of Geology," which was to become the nineteenth century's foundation textbook in the science. Darwin discovered with great pleasure "the wonderful superiority of Lyell's manner of treating geology, compared with that of any other author whose works I had with me or ever afterwards read." Had Lyell told him that the Cape Verde Islands had also been on a voyage—that in a deep geophysical sense they had come from New England—Darwin might have thrown the book overboard.

Even on the best-defined tracks, not everything falls patly into place. There is some granite in New Hampshire that is two hundred million years old—still too young to be part of the Appalachian orogenic story but too old to be explained in terms of the two passing hot spots that left other granites. Possibly the two-hundred-million-year-old rock has something to do with magmas that came up at that time as the crust tore apart to admit the Atlantic. When Great Meteor arrived at the edge of the Canadian Shield, under the present site of Montreal, it presumably made the Monteregian hills, for one of which the city is named. The Monteregian hills are volcanic, but their potassium-argon age disagrees by twenty million years with the date when, by all other calculations, Montreal was over the hot spot—an exception that probes the theory. Morgan attributes the inconsistency to "random things you can't explain" and mentions the possibility of faulty dating. He also says, quite equably, "If the Monteregian hills really don't fit the model, you have to come up with another model."

The hot-spot hypothesis was put forward in the early nineteen-sixties by J. Tuzo Wilson, of the University of Toronto, as a consequence of a stopover in Hawaii and one look at the islands. The situation seemed obvious. James Hutton, on whose eighteenth-century "Theory of the Earth" the science of geology has been built, understood in a general way that great heat from deep sources stirs the actions of the earth ("There has been exerted an extreme degree of heat below the strata formed at the bottom of the sea"), but no one to this day knows exactly how it works. Heat rising from hot spots apparently lubricates the asthenosphere—the layer on which the plates slide. According to theory, the plates would stop moving if the hot spots were not there. Why the hot spots are there in the first place is a question that seeks its own Hutton. For the moment, all Jason Morgan can offer is another shrug and smile. "I don't know," he says. "It must have something to do with the way heat gets out of the lower mantle."

From very deep in the mantle (and perhaps all the way from the core) the heat is thought to rise in a concentrated column, and for this reason is alternatively called a plume. Its surface features are not proof in themselves that they are the product of some plant-stem phenomenon that is (or was) standing in the mantle far below. The chemistry of hot-spot lavas suggests that the rock is coming from below the asthenosphere, but there is no direct evidence of fixed hot spots in the mantle. They exist on inference alone. There is no way to sample the mantle. It can only be sensed—with vibrational waves, with viscosity computations, with thermodynamic calculations of what minerals do at different temperatures and pressures. Sound waves move slowly in soft rock, and some modes of the sound can be stopped completely where the rock is molten. The speed and patterns of seismic waves tell the story of the rock. Seismology is not quite sophisticated enough to look through the earth and count hot spots, but it approaches that capability, and when it gets there hot spots should appear on the screen like downspouts in a summer storm. If they don't, that may be the end of the second-greatest story in the youthful explorations of geological geophysics.

Hot spots seem to be active for roughly a hundred million years. Some of their effects on overriding plates last, of course, longer than they do. If they begin under continents, their initial manifestations at the surface are likely to be flood basalts. Hot-spot tracks have gone forth not only from the flood basalts of the Columbia River and the Serra Geral but in India from the flood basalts of the Deccan Plateau, in South Africa from the flood basalts of the Great Karroo, in East Africa from the flood basalts of the Ethiopian Plateau, in Russia from the flood basalts of the Siberian platform. Flood basalts are what the term implies—geologically fast, and voluminous in their declaration of the presence of a hot spot. In Oregon and Washington, in the middle Miocene, two hundred and fifty thousand cubic kilometres flowed out within three million years. Having achieved the surface in this form, the plume begins to make its track as the plate above slides by, just as Yellowstone, starting off from the flood basalts of Oregon and Washington, stretched out the pathway that has become the Snake River Plain.

An event of the brevity and magnitude of a great basalt flood is an obvious shock to the surface world. "We don't know what flood basalts do to the atmosphere," Morgan remarked one day in 1985, showing me a chronology he had been making of the great flood basalts that not only filled every valley "like water" and killed every creature in areas as large as a million square kilometres but also may have spread around the world lethal effects through the sky. Morgan's time chart of flood basalts matched almost exactly the cycles of death that are currently prominent in the dialogue of mass-extinction theorists, including the flood basalts of the Deccan Plateau, which are contemporaneous with the death of the dinosaurs—the event that is known as the Cretaceous Extinction.

The perforations made by hot spots may be analogous to the perforations in sheets of postage stamps. Plume tracks might weaken the plates through which they pass so that tens of millions of years later the plates would break apart along those lines. Madeira, for example, first drew the line where Greenland broke away from Canada. The Kerguelen Hot Spot, in the Indian Ocean, may have helped India break away from Antarctica. The Crozet Hot Spot, also in the Indian Ocean, seems to have helped Madagascar get away from Africa. In the interior of Gondwanaland, the southern supercontinent of three hundred million years ago, a hot spot punched out the line that is now the north coast of Brazil. The same line is the Gold Coast and Ivory Coast of Africa. The hot spot now stands in the Atlantic as the island St. Helena.

The oldest rocks in Iceland are at the eastern and western extremes of the island, because Iceland is a hot spot whose track comes down from the northwest and at present intersects the Mid-Atlantic Ridge where Europe

and America diverge. Iceland, for the time being, is
spreading with the Atlantic. A hundred million years
ago, the Mt. Etna Hot Spot was under the Ukraine, and
seems to have cleaned off the Ukrainian Shield. A hot
spot has made Ascension Island, on the South American
Plate beside the Mid-Atlantic Ridge, fourteen hundred
miles east of Brazil. It spent a hundred and ten million
years under Africa after starting off from the Bahamas
in early Jurassic time, when the transatlantic crossing
was instantaneous, because there was no Atlantic. The
high-standing Bahamas—eighteen thousand feet above
the Hatteras Abyssal Plain—are defined as a carbonate
platform, its wide shallow seas underlain by limestones
and corals. Morgan says, "I would hope that if you
drilled through them you would end up with basalt."
The Labrador Hot Spot is thought to be "blind"—a hot
spot that has not found a way to drive a plume to the
surface but has nonetheless raised the terrain. This
would account for the otherwise unaccountable altitudes of Labrador, not to mention more cleansing of the
Canadian Shield. The Guiana Shield is also thought to
lie above a blind hot spot, which has lifted the country
and produced, among other things, the world's highest
falls—a plume of water twenty times the height of
Niagara.

Bermuda is the last edifice of a faint but evident
hot spot, which underlies the ocean crust east of the
present islands. The domal swell of the seafloor is classic—like Hawaii's, a thousand kilometres wide. (Under
continents, upwelled masses analogous to the Bermu-
dian and Hawaiian swells can be shown by satellite
measurements of gravity anomalies.) Bermuda has not

been active for thirty million years, but its track can be extrapolated westward in conformity with the track of Great Meteor and the well-established motions of the North American Plate. Seen in its former contexts, Bermuda proves to be a good bit less interesting for where it is now than for where it has been. If you could somehow look into the side of the American continent from Georgia to Virginia, you would see a great suite of Cretaceous strata dipping north and south, descending like a rooftop from an apex at Cape Fear. Something lifted up that arch, and, as one can readily discern from the stratigraphy and structure, whatever did the lifting did it in Paleocene time. Since the Paleocene, the North American Plate has moved the exact distance from Bermuda to Cape Fear.

Bermuda came through there like a train coming out of a tunnel. Or so it would appear. In the Campanian age of late Cretaceous time, when Great Meteor was in mid-Atlantic, Bermuda was under the Great Smoky Mountains. The Appalachian system consists of parallel bands of kindred geology sinuously winding from Newfoundland to Alabama, where they disappear under the sediments of the Gulf Coastal Plain. Why this long ropy package would stand high in two places and sink low in others is not explained by plate tectonics. It can be explained by hot spots. Great Meteor and Cape Verde seem to have lifted New England's high mountains, Bermuda the Smokies. Uplift accelerates erosion. The rock of the Permian period—the last chapter in the Appalachian mountain-building story—has been removed everywhere in eastern America except in West Virginia and nearby parts of Ohio and Pennsylvania, halfway between the hot-spot tracks, halfway between New Hampshire and North Carolina. Because plate motions have shifted over time, the tracks of all hot spots, ancient and modern, form a palimpsest on the face of the earth. Untouched areas between lines often prove to be continental basins—the Michigan Basin, the Illinois Basin, the Mississippi Embayment, the Williston Basin—while the rims of the basins are structural arches lined up on the tracks of the hot spots. Morgan thinks the large continental basins may have been created when hot spots elevated the edges. The Great Meteor track runs between the Hudson Bay Basin and the Michigan Basin. A Paleozoic hot spot seems to have made the Kankakee Arch, which separates the Michigan and Illinois Basins. The Bermuda track runs between the Illinois Basin and the Mississippi Embayment. "Every basin gets missed," comments Morgan, with his hand on a map. "I don't think that's a coincidence."

Bermuda made the Nashville Dome. It lifted the Ozark Plateau, in middle Cretaceous time. "How much erodes off the top when a hot spot lifts something up depends on the durability of what's there," Morgan goes on. "If it's coastal mush, or Mississippi River mush, it goes quickly and in great volume. If it's quartzite, it resists. The resistant stuff stands up higher." Much of a hot spot's energy is expended in thinning the plate above it. Where the plate is already thin, most of the energy will appear at the surface in outpourings such as lava flows. When a plume has to come up through thick old craton, it makes kimberlites, carbonatites, gas-rich blowouts. The plumes may express themselves as diatremes, the extremely focussed volcanic events that bring diamonds out of the mantle and explode them into the air at Mach 2. The conduits are so narrow they are called pipes, and the rock left inside them after the explosion is kimberlite. When Bermuda was under Kansas, it sent up the Riley County kimberlites. For many years, these diamond pipes were described as crypto-volcanic structures, meaning that nobody knew what they were. Later, they were thought to be meteorite strikes. In 1975, in Riley County, a hole was drilled with a tungsten-carbide bit that could smoothly cut its way through anything but diamonds. It went down sixty feet, where all penetration ceased. The bit was pulled. It was grooved and scarred. There are "meteor impacts" along the Bermuda track in Tennessee, southern Kentucky, and Missouri. Morgan thinks they are diatremes, or, as he puts it, "hot-spot blasts." They lie in a matrix of Paleozoic rock. If in fact they are meteor impacts, the hot spot would have lifted up the country and caused the erosion that exposed them to view. In Morgan's summation, "the thing works for me either way."

When Bermuda was under Wyoming, in Neocomian time, the Rockies did not exist, but the magmas of the Idaho batholith had recently come in, a short distance up the track. When Bermuda was under the state of Washington, the state of Washington was blue ocean. If the track is followed back to two hundred million years, Bermuda seems to have been under Yakutat, Alaska. Hearing most of this for the first time at a colloquium in Princeton, a graduate student said, "This is like playing chess without the rules."

During the past twenty million years, the region that we like to call the Old West is thought to have

been passing over not one but two hot spots, which
have done much to affect the appearance of the whole
terrain. The other one is less intense than Yellowstone,
and is at present centered under Raton, New Mexico.
Volcanoes are at the surface there. The Raton plume has
lifted the Texas panhandle, the southern Colorado high
plains. Its easternmost lava flow is in western Oklahoma.
Its track, parallel to Yellowstone's, includes the Jemez
Caldera, above Los Alamos, and may have begun in the
Pacific. To the question "What lifted the Colorado
Plateau, the Great Plains, and the Rocky Mountain platform?" the answer given by this theory is "The plumes
of Raton and Yellowstone." As Utah and Nevada crossed
the hot spots, the plumes are thought to have initiated
the extensional faulting that has separated the sites of
Reno and Salt Lake City by sixty miles in eight million
years, breaking the earth into fault blocks and creating
the physiographic province of the Basin and Range.
Work done in the rock-dating laboratory of Richard
Armstrong, a geochemist and geochronologist at the
University of British Columbia, shows that Basin and
Range faulting began at the western extreme of the
region and moved eastward at a general rate of twenty-
eight miles per million years—a frame commensurate
in time and space with the continent's progress over
the hot spots now positioned under Yellowstone and
Raton. The Tetons began to rise eight million years ago
and are clearly not products of the Laramide Orogeny.
They are a result of extensional faulting, and conform
to hot-spot theory as the easternmost expression of the
Basin and Range. The Colorado Plateau lies between
the two hot-spot tracks, and Morgan believes that their

combined influence is what lifted it, setting up the hydraulic energy that has etched out the canyonlands. How the plateau avoided the rifting and extension that went on all around it—why it, too, did not break into blocks—is a question that leaves him baffled. That the two hot spots, at any rate, are progressively lifting the country is a point reinforced by a remarkable observation: a line drawn between them is the Continental Divide.

Inevitably, it has been suggested that someday North America may split apart along the Yellowstone perforations of the Snake River Plain. "That gives me a caution," says David Love. "I think there are some problems there. I have a feeling that the hot-spot ideas have been somewhat enlarged beyond the facts. The term itself probably means different things to different people. To me, a hot spot is an area of abnormally high temperature gradients, so high that it can be interpreted as having an igneous mush down below. In the Snake River Plain, the volcanics do get older east to west— in a broad sense, yes. But when you get down to details you get down to discrepancies. We don't know all the ages we should, on the various sets of volcanics. We need to learn them, and plot them up in geographic and time perspective. We will—but to my satisfaction we have not, yet. I would like to see a lot more regional information. In northwest Wyoming, volcanism began in the early Eocene, fifty-two million years ago. You got the Absaroka volcanic centers. Volcanic debris from them was spread by water and wind across the Wind River Basin, the Green River Basin. Then what hap-

pened? Everything went blah. The Yellowstone-
Absaroka hot spot abruptly terminated at the end of
Eocene time. Where the hell did that hot spot go?
Twenty-five to thirty million years later, it was reacti-
vated in the same place. What was that plume doing
for all those millions of years? How do you reactivate a
plume? We need answers to this sort of thing, and we
don't have them. If the plume theory is correct, you've
got to answer those questions."

The hail over the interstate turned to snow, and we
passed a Consolidated Freightways tandem trailer lying
off the shoulder with twenty-six wheels in the air—apparently overturned (a day or two before) by the wind.
Abruptly, the weather changed, and we climbed the
Rock Springs Uplift under blue-and-white marble skies.
As we moved on to Green River and Evanston—across
lake deposits and badlands, and up the western overthrust—the sun was with us to the end of Wyoming. On
the state line was a flock of seagulls, in the slow lane,
unperturbed, emblematically announcing Utah—these
birds that saved the Mormons. Mormon traffic, heading
home, did not seem intent on returning the favor.

If Wyoming can be said to have been acupunctured
for energy, nowhere was this so variously evident as in
the southwestern quadrant of the state, from the new

coalfields near Rock Springs to the new oil fields of the Overthrust Belt, not to mention experimental attempts to extract petroleum from Eocene lacustrine shale, which—in that corner of Wyoming and adjacent parts of Colorado and Utah—contains more oil than all the rock of Saudi Arabia. More than the Union Pacific was after such provender now. "We are at the mercy of the East Coast and West Coast establishments," Love said. "It's been called energy colonization." And while we traversed the region, with scene after scene returning us to this theme, his reactions were not always predictable. There were moments that emphasized the scientist in him, others that brought out the fly-at-it-folks discoverer of resources, and others that brought forth a vigorous environmentalist, conserving his native ground, fulminating in the face of effronteries to humanity and the earth. Love is a prospector in the name of the people, who looks for the wealth in exploitable rock. He is also a pure scientist, who will follow his instincts wherever they lead. And he is a frequent public lecturer who turns over every honorarium he receives to organizations like the Teton Science School and *High Country News*, whose charter is to understand the environment in order to defend it. Thus, he carries within himself the whole spectrum of tensions that have accompanied the rise of the environmental movement. He carries within himself some of the central paradoxes of his time. Among environmentalists, he seems to me to be a good deal less lopsided than many, although beset by contradictory interests, like the society he serves. He cares passionately about Wyoming. It may be acupunctured for energy, but it is still Wyoming, and only words and images, in their inevitable concentration, can effectively clutter its space: a space so great that you can stand on a hilltop and see not only what Jim Bridger saw but also—through dimming tracts of time—what no one saw.

The Rock Springs Uplift, like the Rawlins Uplift, is a minor product of the Laramide Revolution, a hump in the terrain which did not keep rising as mountains. There was "red dog"—red clinker beds—in low cuts beside the road. When a patch of coal is ignited by lightning or by spontaneous combustion, it will oxidize the rock above it, turning it red. The sight of clinker is a sign of coal. Love said that this clinker was radioactive. Like coal, it was adept at picking up leached uranium. As the cuts became higher, we could see in the way they had been blasted the types of rock they contained. Where the cuts were nearly vertical, the rock was competent sandstone. Where the backslope angle was low, you knew you were looking at shale. Cuts that went up from the road through sandstone, then shale, then more sandstone, had the profiles of flying buttresses, firmly rising to their catch points, where they came to the natural ground. The shallower the slope, the softer the rock. The shallowest were streaked with coal.

At Point of Rocks, a hamlet from the stagecoach era, was a long roadcut forty metres high, exposing the massive sands of a big-river delta, built out from rising Rockies at the start of the Laramide Orogeny into the retreating sea. We left the interstate there and went north on a five-mile road with no outlet, which followed the flank of the Rock Springs Uplift and soon

curved into a sweeping view: east over pastel buttes
into the sheep country of the Great Divide Basin, and
north to the white Wind Rivers over Steamboat Mountain and the Leucite Hills (magmatic flows and intrusions, of Pleistocene time), across sixty miles of
barchan dunes, and, in the foreground—in isolation
in the desert—the tallest building in Wyoming. This
was Jim Bridger, a coal-fired steam electric plant,
built in the middle nineteen-seventies, with a generating capacity of two million kilowatts—four times
what is needed to meet the demands of Wyoming.
Twenty-four stories high, the big building was more
than twice as tall as the Federal Center in Cheyenne,
which is higher than Wyoming's capitol dome. Rising
beside the generating plant were four freestanding col-
umnar chimneys so tall that they were obscured in
cumulus from the cooling towers, which swirled and
billowed and from time to time parted to reveal the
summits of the chimneys, five hundred feet in the air.
"This place is smoking the hell out of the country," Love
said. "The wind blows a plume of corruption. In cold
weather, sulphuric acid precipitates as a yellow cloud.
It's not so good for people, or for vegetation. Whenever
I think of this plant, I feel sadness and frustration. We
could have got baseline data on air and water quality
before the plant was built, and we muffed it." He blames
himself, although at that time he had arsenic poisoning
from springwater in the backcountry and was sick for
many months.

The idea behind Jim Bridger was to ship energy
out of Wyoming in wires instead of railway gondolas.

Ballerina towers, with electric drapery on their out-
stretched arms, ran from point to point to the end of
perspective, relieving pressure on the Oregon-Idaho
grid. The coal was in the Fort Union formation—in a
sense, the bottom layer of modern time. Locally, it was
the basal rock of the Cenozoic, the first formation after
the Cretaceous Extinction—when the big animals were
gone, but not their woods and vegetal swamps. Wyoming had drifted a few hundred miles farther north
than it is now, and around the low swamplands were
rising forests of oak, elm, and pine. The terrain was near
sea level. Mountains had begun to stir—Uintas, Wind
Rivers, Owl Creeks, Medicine Bows—and off their
young slopes they shed the Fort Union, its muds bury-
ing the compiled vegetation, cutting off oxygen, preserving the carbon. As the mountains themselves became
buried, the fallen vegetation in the thickening basins
was ever more covered as well, to depths and pressures
that caused it to become a soft and flaky sub-bituminous
low-rent grade of coal, a nonetheless combustible low-
sulphur coal. With the Exhumation of the Rockies, nature, in the form of wind and water, worked its way
down toward this coal. By the middle nineteen-
seventies, nature had removed a mile of overburden,
and had only sixty feet to go. At that point, something
called the Marion 8200, an eight-million-pound landship
also known as a walking dragline, took over the job.

The machine was so big it had to be assembled on
the site—a procedure that required fourteen months.
Now working within a mile or two of the generating
plant, it could swing its four-chord deep-section boom

and touch any spot in six acres, its bucket biting, typically, a hundred tons of rock, and dumping it to one
side. The 8200 had dug a box canyon, its walls of solid
coal about thirty feet thick. The inside of the machine
was painted Navy gray, and had non-skid deck surfaces,
thick steel bulkheads, handrails, and oval doors that
looked watertight. They led from compartment to compartment, and eventually into the air-conditioned sanc-
tum of Centralized Power Control, where, lined up in
ranks, were electric motors. The foremost irony of this
machine was that it was far too large and powerful to
operate on diesel engines. Although the chassis was nine
stories high, it could not begin to contain enough diesels
to make the machine work. Only electric motors are
compact enough. Out the back of the machine, like the
tail of a four-thousand-ton rat, ran a huge black cable,
through gully and gulch, over hill and draw, to the generating plant—whose No. 1 customer was the big
machine.

Once every couple of hours, the 8200 walked—
raised itself up on its pontoonlike shoes and awkwardly
lurched backward seven feet, so traumatically com-
pressing the dirt it landed on that smoke squirted out
the sides and the ground became instant slate. This
machine—with its crowned splines, its precise driveline
mating, its shop-lapped helical gears, its ball-swivel
mounting of the boom-point sheaves, its anti-tightline
devices and walking-shoe position indicators—had un-
surprisingly attracted the attention of Russian engineers,
who came in a large committee to see Jim Bridger, because they were about to build twenty-five similar gen-

erating stations in one relatively concentrated area of Siberia, which, they confided, closely resembled Sweetwater County, Wyoming.

This strip mine, no less than an erupting volcano, was a point in the world where geologic time and human time had intersected. Ordinarily, the close relationship between the two is masked: human time, full of beepers and board meetings, sirens and Senate caucuses, all happening in microtemporal units that physicists call picoseconds; geologic time, with its forty-six hundred million years, delivering a message that living creatures prefer to return unopened to the sender. In this place, though, geology had come up out of its depths to join the present world, and, as Love would put it, all hell had broken loose. "How people look at it depends on whose ox is being gored," he said. "If you're in a brownout, you think it's great. If you're downwind, you don't. Wyoming's ox is being gored."

When the Bridger operation was under construction, hundreds of tents and trailers lined most of the five miles of the spur road to the site—an "impact" that ultimately shifted to Rock Springs, thirty miles away, and Superior, and other small towns in the region. Populations doubled during the coal rush, which was close in time to the booms in trona mining and oil. Even after the booms had settled down, twenty-eight per cent of the people of Wyoming were living in mobile homes. During the construction of Jim Bridger, Rock Springs, especially, became a heavy-duty town, attracting people with no strong attachments elsewhere who came into the country in pickups painted with flames. With its bar fights and prostitutes, it was wild frontier territory, or seemed so to almost everyone but David Love. "Fights were once fights," he commented. "Now the fight starts and your friends hold you back while you throw insults." Cars were stripped of anything that would come off. Pushers arrived with every kind of substance that could stun the human brain. A McDonald's sprang up, of course, decorated with archaic rifles, with plastic cattle brands lighted from the inside, with romantic paintings of Western gunfights—horses rearing under blazing pistols on dusty streets lined with false-fronted stores. A Rock Springs policeman shot another Rock Springs policeman at point-blank range and later explained in court that he had sensed that his colleague was about to kill him. How was that again? The defendant said, "When a man has the urge to kill, you can see it in his eyes." The jury saw it that way, too. Not guilty. Some people in Sweetwater County seemed to be of the opinion that the dead policeman needed killing.

Love's son Charlie, who lives and teaches in Rock Springs, once told us that the community's underworld connections were "only at the hoodlum level." He explained, "The petty gangsters here aren't intelligent enough for the Mafia to want to contact. You can't make silk purses out of sows' ears."

The number of cowboys in Wyoming dropped from six thousand to four thousand as they rushed into town to join the boom, disregarding the needed ratio of one man per thousand head of cattle. In desperation for help at branding time, calving time, and haying time,

ranchers had to go to the nearest oil rig and beg the
roughnecks to moonlight.

For a steam-driven water-cooled power plant, this
one seemed to have a remarkably absent feature. It
seemed to be missing a river. The brown surrounding
landscape was a craquelure of dry gulches. In one of
them, though—a desiccated arroyo called Dead Man
Draw—was a seventy-five-acre lake, fringed with life
rings, boats, and barbecue grills. At the rate of twenty-
one thousand gallons a minute, Jim Bridger was sucking water from the Green River, forty miles to the
west. To cool an even drier power station, some hundreds of miles away in northeast Wyoming, a proposal
had been made to pump Green River water over the
Continental Divide to the Sweetwater River, which
runs into the North Platte, from which the water would
be pumped over a lesser divide and into the Powder
River Basin. Love said, "That would destroy the whole
Sweetwater regimen, destroy the Platte, and destroy the
Powder River, all for coal in the Powder River Basin—a
slurry pipeline or something of the sort. It's very much
on the books. If they go in for the gasification of coal,
they're going to need it. It's known as the trans-basin
diversion of the Green River. The water has fluorine
in it. Wherever it gets into the ground, it can pollute the

water table in ten to fifteen years. The river also picks up sodium from trona. In the town of Green River, the sodium in the drinking water greatly exceeds E.P.A. standards. If they decide to pipe the water over the Continental Divide, water quality could be lowered in the Powder River Basin to the point of needing a desalinization plant."

We moved on toward Green River, where the most spectacular suite of roadcuts and rock exposures anywhere on Interstate 80 contained in its sediments the history of these evils. Dark mountains, spread low across the horizon, might have been a storm coming on—and in a sense they were, or had been. They were the Overthrust Belt, cumulate from the west. Looking north to the even more distant Gros Ventres and Wind Rivers, and south to the high cirques of the Uintas, we were encompassing in a wide glance about sixteen thousand square miles of land, much of it so dry, stacked flat like crumbling hardtack, that only a geologist could absorb such a scene and see in it a lake that would rank seventh in the world.

In the Eocene, when the lake existed, the appearance of North America approached its present form. The journey from New York to Paris may have been eight hundred miles shorter than it is now, but the North Atlantic was a maturing ocean. The Appalachians were much higher. There were no Great Lakes. There was no westerly rise to the Great Plains. The foreland ranges of the Rockies had pooched out from their sea-level platform, and west-running rivers were flowing around them to pool against the overthrust mountains. In California there was no Sierra, in Nevada and Utah no mountains of the Basin and Range—only moist gentle country coming in from the Pacific Coast. This Eocene time line, drawn from either end of the continent, would have converged in western Wyoming in something comparable to the Sea of Azov. A hundred and fifty miles long, a hundred miles wide, it was larger by far than Erie, larger than Lake Tanganyika, larger than Great Bear. It was two hundred times Lake Maggiore. It had no name until a century ago, when a geologist called it Lake Gosiute.

Lakes are so ephemeral that they are seldom developed in the geologic record. They are places where rivers bulge, as a temporary consequence of topography. Lakes fill in, drain themselves, or just evaporate and disappear. They don't last. The Great Lakes are less than twenty thousand years old. The Great Salt Lake is less than twenty thousand years old. When Lake Gosiute took in the finishing touch of sediment that ended its life, it was eight million years old.

West of Rock Springs, we came to an escarpment known as White Mountain, standing a thousand feet above the valley of Killpecker Creek. In no tectonic sense was this a true mountain—a folded-and-faulted, volcanic, or overthrust mountain. This was just a Catskill, a Pocono, a water-sliced segment of layered flat rock, a geological piece of cake. In fact, it was the bed of Lake Gosiute, and contained almost all of the eight million years. Apparently, the initial freshwater lake eventually shrank, became bitter and saline, and intermittently may have gone dry. Later, as the climate remoistened, water again filled the basin, and the lake reached its greatest size. As we looked at White Mountain, we could see these phases. It was the dry, salt-lake interval in the middle—straw and hay pastels so pale they were nearly white—that had given the bluff its name. The streams that had opened it to view were lying at its base. Killpecker Creek (full of saltpeter) flowed into Bitter Creek, and that soon joined the Green River.

Down the road a couple of miles was a pair of tunnels—snake eyes in the lakebed. They were one of only three sets of tunnels on Interstate 80 between New York and San Francisco, but they had to be there in the nose of White Mountain, or the interstate, flexing left, would destroy the town of Green River. Tower sandstone stood on the ridgeline in castellated buttes. With each mile, they increased in number, like buildings on the outskirts of a city. Off to the left was the island from which the geologist John Wesley Powell—seven years before the battle of the Little Bighorn—set off in a flotilla of dinghies to follow the Green River into its master stream, and to survive the preeminent rapids of North America on the first known voyage through the Grand Canyon. A huge sandstone broch stood in brown shale above the tunnels, which penetrated the lakebed's saline phase. If the Great Salt Lake, which has been freshening in recent years, should follow the biography of Gosiute, it will swell up to the size of Huron and, as it once did, spill to the north over Red Rock Pass and pour down the Snake River Plain. Inundating much of Utah and some of Nevada, it will send the Mormons back to New York, saying, "That was the place."

We burst into the light at the western end of the tunnel among concentrated stands of lofted redoubts, a

garden of buttes, and huge walls of flat strata in roadcuts and rivercuts extending a full mile. William Henry
Jackson photographed this scene, in 1870, for the Hayden Survey. The buttes have been given names like Tollgate Rock, Teakettle Rock, Sugar Bowl Rock, Giant's
Thumb. Love said there were Indian petroglyphs on
Tollgate Rock but they were far too high to see. "You'd
have to be a mountain goat to get up there," he went on,
and scarce had he uttered the words when a figure
white as gypsum appeared on a cone of talus at the
base of the Tower sandstone, close by the petroglyphs,
its head in motionless silhouette. "You can tell people
just to look for that goat if they want to see where the
petroglyphs are," he advised me. "They can always find
the petroglyphs by looking for the goat."

About halfway up White Mountain was a layer of
sandstone that happened to be phosphatic and contained uranium. Love said he knew this because he had
discovered the uranium. Non-marine phosphate, largely
unknown elsewhere in the world, was one of the many
legacies of this strange vanished lake. A few miles back,
the uraniferous phosphatic sandstone had formed a low
ridge in the path of the interstate, which cut straight
through it, dosing all drivers with a few milliroentgens
to keep them awake.

He also remarked that the sedimentary story was
reflecting a lot of tectonic history. You could see the
orchestration of the mountain ranges by reading backward through the layers of sediment. For example, from
the age and position of the sedimentary rock derived
from the Uinta Mountains and the Wind River Range
you could see the Wind Rivers developing first.

[*191*]

At all moments in the history of Lake Gosiute, it was replete with organic life, from the foul clouds of brine flies that obscured its salty flats to the twelve-foot crocodiles and forty-pound gars in the waters at their widest reach. For this was Wyoming in the Eocene, and in the lake at varying times were ictalurid catfish, bowfins, dogfish, bony tongues, donkey faces, stingrays, herring. The American Museum of Natural History has a whole Gosiute trout perch in the act of swallowing a herring, recording in its violence two or three seconds from forty-six million years ago. In the museum's worldwide vertebrate collection, roughly one fossil in five comes from Wyoming, and a high percentage of those are from Gosiute and neighboring lakes. Around the shores were red roses and climbing ferns, hibiscus and soapberries, balloon vines, goldenrain. The trees would generally have been recognizable as well: pines, palms, redwoods, poplars, sycamores, cypresses, maples, willows, oaks. There were water striders, plant hoppers, snout beetles, crickets. The air was full of frigate birds. Dense beds of algae matted the shallows. In all phases through the eight million years, quantities of organic material mixed with the accumulating sediments and are preserved with them today in the form of oil shale. On the far side of the Uinta Mountains was another great lake, reaching from western Colorado well into Utah. Lake Uinta, as it has come to be called, and Lake Gosiute and several smaller lakes left in their shales a potential oil reserve recently estimated at about one and a half trillion barrels. This is the world's largest deposit of hydrocarbons. It is actually nine times the amount of crude oil under Saudi Arabia, and about ten times as much oil as has so far been pumped from American rock.

Distinct in the long suite of cuts at Green River were the so-called mahogany ledges, where oil shale is particularly rich. They looked less like wood than like bluish-white slabs of thinly bedded slate. Oil shale always weathers bluish white but is dark inside, and grainy like wood. The thinner the laminae the higher the ratio of organic material. The richest of the dark oleaginous flakes—each representing the sedimentation of one year—were fifteen-thousandths of a millimetre thick. Love dropped some hydrochloric acid on the rock, and the acid beaded up like an arching cat. "It's actually kerogen," he said. "It converts to high-paraffin oil. It's not like Pennsylvania crude."

To mining engineers, oil shale had presented an as yet unsolved and completely unambiguous problem: how to remove the shale without destroying the face of the earth. So far, three principal methods had been considered. One was to strip-mine it, crush it, separate the oil, then smooth out the tailings—a process that could result in the absolute rearrangement of twenty-five thousand square miles. Another was to go underground, excavate a percentage of the rock, and refill the caverns with tailings. That was known as the "modified in situ" approach. And finally someone thought of drilling a hole, pumping in propane, and starting a fire. The heat would cause liquid oil to run out of the shale. The oil could be forced up through another well before the fire destroyed it. A burn would not, like a clinker fire, continue indefinitely. If oxygen was not fed to the flames, they would die. This was known as "true in situ" mining; and there in White Mountain, a few miles away, the federal government had been perfecting the technique. The experiments thus far had brought down the recovery cost to a million dollars a barrel. In Cheyenne one time, I saw a Peter Pan Crunchy Peanut Butter jar filled with such oil. It looked and smelled like the contents of a long unemptied spittoon.

The one-and-a-half-trillion-barrel estimate was somewhat extravagant, because it included every last drop—referring, as it did, to all shale with any content of kerogen. In the richer rock—in the shales that contained from twenty-five to sixty-five gallons of oil per ton—were no more than six hundred billion barrels. That would do. That was more petroleum in place than all the petroleum produced in the world to date. Love remarked that oil shale had been "trumpeted to the skies" but, with the energy crisis in perigee, both government and industry were losing interest and pulling out. Temporarily pulling out. Sooner or later, people were going to want that shale.

For Lake Gosiute to have lasted so long in a mountain setting, Love said, an amazing delicacy of crustal balance was required. As the lakebed thickened, it had to subside at an appropriate rate if it was to continue to hold water and accept sediment. Gosiute sediments average about half a mile, top to bottom. The oil is at

all levels. The evaporite phase, in the middle, reports
a Gosiute of dense and complex brine that was sur-
rounded by mud flats sickeningly attended by the hum
of a trillion flies. Trona—sodium sesquicarbonate—precipitated out of the brine in concentrations rare in the
world. It was discovered in 1938, but the boom did not
begin until the sixties. We tasted some salty crystals in
the rock at Green River, in beds that dipped west and
pointed into the ground toward mines. Trona is an
important component of ceramics and textiles, pulp
and paper, iron and steel, and, most especially, glass.
Love commented that more than two tons of trona had
been going into the Green River every day merely from
the washing of freight cars—and that was a lot of sodium. The Wyoming Department of Environmental
Quality had put a stop to the practice. He said there had
been a brewery in Green River that drew its water from
a well drilled to trona. The beer had a head like a stom-
ach tablet. A few miles south of us were the headwaters
of the reservoir that covered Flaming Gorge. Before
the federal Bureau of Reclamation built a dam there,
Flaming Gorge was one of the scenic climaxes of the
American West—a seven-hundred-foot canyon in arching Triassic red beds so bright they did indeed suggest
flame. Afterward, not much was left but the hiss, and
an eyebrow of rock above the water. The reservoir
stilled fifty miles of river. Some of the high water penetrates beds of trona. When the reservoir drops, dissolved
trona comes out of the rock and drips into the reservoir.
When water rises again, it goes back into the rock for
more trona. Love said that Lake Powell and Lake
Mead—reservoirs downstream—were turning into

chemical lakes as a result. "And a lot of it winds up with the poor farmers in Mexico," he said. "We are going to have to desalinate their water." Some miles along the interstate, when we crossed the Blacks Fork River, we would see alkali deposits lying in the floodplain like dried white scum. On both sides of the road were abandoned farmhouses, abandoned barns, their darkly weathered boards warping away from empty structures out of plumb. The river precipitates and the abandoned farms were not unrelated. This was the Lyman irrigation project, Love explained—a conception of the Bureau of Reclamation, an attempt to make southwestern Wyoming competitive with Wisconsin. The Blacks Fork River was dammed in 1971, and its waters were used to soak the land. The land became whiter than a bleached femur. It still appeared to be covered with light snow. "Alkali sours the land," Love said. "The drainage here is just too poor to flush it out. Imagine the sodium those farmers drank in their water."

Meanwhile, west of Green River, a tall incongruous chimney seemed to rise up out of the range, streaming a white plume downwind. Below the chimney, but hidden by the roll of the land, was a trona refinery, and, below the refinery, a mine. I had gone down into it one winter day half a dozen months before, and I now remarked that the people there had told me that the white cloud issuing from the chimney was pure steam.

"It goes clear across the state," Love said. "That's pretty durable for steam."

He said that fluorine, among other things, was coming out of the refinery with the steam. Settling downwind, it could cause fluorosis. He thought it might be

[*196*]

damaging forests in the Wind River Range. The afternoon sky was cloudless but not exactly clear. "The haze
you see is the trona haze that goes across Wyoming,"
he continued. "We never used to have this. You could
clearly see distant mountains on any average day."

Trona is about as hard as a fingernail, and much of
it looks like maple sugar or honey-colored butter crunch.
I remembered drinking coffee at a picnic table nine
hundred feet below us, in a twilighted Kafkan dusty
world where dynamite provoked reverberate thunders
that moved from room to room and eventually clapped
themselves. Chain saws with bars ten feet long sliced
into the rock to define the next blast. Stickers on lunch
pails said:

Don't Tempt Fate.

I Have Met the A-O Dust Demon.

When Escape Is Cut Off: 1. Barricade   2. Listen for 3
Shots   3. Signal by Pounding Hard 10 Times   4. Rest 15
Minutes Then Repeat Signal Until   5. You Hear 5 Shots,
Which Means You Are Located and Help Is on the Way.

"The Southeast is the stroke and hypertension belt
of the United States," Love was saying. "That is blamed
on sodium, including sodium in the water. We're not
far behind. Perhaps we can overtake them."

By the Gros Ventre River near Crystal Creek, some
years ago, Love noticed horses eating the Cloverly formation—putting their noses right on the outcrop and
slurping up nodules of soft Cretaceous lime. He could
guess where the horses had come from. They were from
Cora, near Pinedale, at the western base of the Wind

River Mountains. When the Wind Rivers came rising up during the Laramide Revolution and moved a few miles west, they completely covered the only limestone in the region. As a result, he said, it is not unusual for a college freshman who has grown up in Pinedale to require false teeth. Pinedale has one of the two or three highest records of dental decay in Wyoming. Pinedale is to caries as Savannah is to coronary thrombosis, in each case for a geological reason.

He said that somewhere in limbo on the industrial drawing board was a geothermal project that would mine the hot groundwaters of the Island Park Caldera, southwest of Yellowstone. The question uppermost in many people's minds seemed to be: What would happen to Old Faithful and other Yellowstone geysers? In New Zealand, when the government tapped the fifth-largest geyser field in the world for geothermal energy the Karapiti Blowhole shut down as promptly as if a hand had turned a valve. A geyser field in Nevada once rivalled Yellowstone's—until 1961, when geothermal well-drilling killed the Nevada geysers. Old Faithful was having trouble enough without help from the hand of man. For a century, and who knows how much longer, Old Faithful had erupted at intervals averaging seventy minutes, but in 1959 an earthquake centered nearby in Montana slowed the geyser down. Additional earthquakes in 1975 and 1983 caused Old Faithful to become so erratic that visitors complained. Constructed around the geyser is something that resembles a stadium, where crowds collect in bleachers and expect Old Faithful to be faithful: "to play," as hydrologists put it—to burst in timely fashion from its fissures, like a cuckoo clock made of water and steam. Frustrated travellers, sometimes clapping their hands in unison, seemed to be calling on the National Park Service to repair the geyser. A scientist confronted with these facts could only shrug, make observations, and formulate a law: *The volume of the complaints varies inversely with the number of miles per gallon attained by the vehicles that bring people to the park.*

Love, who has made a subspecialty of the medical effects of geology, had other matters on his mind. In public lectures and in meetings with United States senators, he asked what consideration was being given to radioactive water from geothermal wells, which would be released into the Snake River through Henrys Fork and carried a thousand miles downstream. After all, radioactive water was known from Crawfish Creek, Polecat Creek, and Huckleberry Hot Springs, not to mention the Pitchstone Plateau. On the Pitchstone Plateau were colonies of radioactive plants, and radioactive animals that had eaten the plants: gophers, mice, and squirrels with so much radium in them that their bodies could be placed on photographic paper and they would take their own pictures. A senator answered the question, saying, "No one has brought that up."

￼At the Overthrust Belt—glamorized of late for its fresh new yields of regular and unleaded—we moved up in topography, down in time, because the great thrust sheets are older than the rock on which they came to rest. The first high ridge was Cretaceous in age, and we left the interstate to climb it, on an extremely steep double-rutted dirt road that led to a mountain valley—a so-called strike valley, of a type that will form where upturned strata angle into the sky and a section is softer than those that flank it. The high valley was fringed with junipers, and, from its eastern rim, presented a view that would impress an astronaut. To look from left to right was to see a hundred and fifty miles, from the Uinta Mountains to the Wind River Range, interspersed with badlands. The badlands were late-Eocene river muds and sands chaotically distributed on top of the filled-in lake, and now being further strewn about at the whim of cloudbursts.

In the center of the high swale were the silvery-gray remains of hundreds of cut trees, which had been dragged into the open and arranged as a fence in kidney-bean shape, all but enclosing about fifteen acres. Vaguely, they formed a double corral, with an aperture in one place only, and had apparently been used—for uncounted years—to trap antelopes. Antelopes don't climb fences, as people fond of roast pronghorn discovered centuries ago. Love's son Charlie, the professor of anthropology and geology at Western Wyoming Community College, knew of the trap and had thought out the strategies by which it was effective. His father expressed pride in Charlie for "thinking as intelligently as the aborigines." The high valley held fast an aesthetic silence—a silence reminiscent of the Basin and Range, a silence equal to the winter Yukon. About the only sign of humankind was the antelope trap. This was the Overthrust Belt as it had appeared before white people —thinking intelligently but not like the aborigines— mapped the terrain, modelled its structure, and went after what lay beneath it. There were mountain bluebells and salt sage in the valley, ground phlox and prickly pear. Love reached down and plucked up a plant and asked if I knew what it was. It looked familiar, and I said, "Wild onion."

He said, "It's death camas. It brings death quickly. It killed many pioneer children. They thought it looked like wild onion."

Suddenly, the great silence was smashed by running gunfire as two four-wheel-drive vehicles, each with a lone rider, appeared over the western rim and thundered up the valley, leaving behind them puffs of blue smoke. They disappeared to the north, still shooting. This was boom country now, however temporarily— another world of pickups painted with flames. It had been described in journals as "the hottest oil-and-gas province in North America"—a phrase in which Love found bemusement and irony, because for three-quarters of a century the hottest oil-and-gas province in North America had been lying there neglected.

"This region was written up in 1907 as containing possible oil fields," Love said. "They're 'finding' them now. That 1907 paper, by A. C. Veatch, of the U.S.G.S., was simply ignored. Until 1975, people said there was no oil in the thrust belt. Now it's the hot area. Veatch did his work in the part of the thrust belt that straddles

I-80. He said oil should be there, and he said where. His paper is a classic. That it was ignored shows the myopia of oil companies, and of geologists in general. The La Barge oil field, in the Green River Basin off the edge of the thrust belt, was discovered in 1924. Twenty years later it became evident that the La Barge field was producing more oil than the structure could contain. The oil was migrating into it from the thrust belt. The evidence was there before us, and we didn't see it. We talked about it. We wondered why. Now the margins of basins have become new frontiers for oil. Anywhere that mountains have overridden a basin, there are likely to be Cretaceous and Paleocene rocks below, quite possibly with oil and gas. The Moncrief oil company drilled through nine thousand feet of granite at Arminto and into Cretaceous rocks and got the god-damnedest field you ever saw."

On I-80 to the end of Wyoming, we moved among the drilling rigs and pump jacks of some of the most productive fields you ever saw. Love said, "These rigs are not damaging the landscape very much. It isn't all or nothing. It doesn't have to be." I remembered a time when we had gazed down into the Precambrian meta-sediments of a taconite mine off the southern tip of the Wind River Range. It was an open pit, square, more than a mile on a side. I asked him how he felt about a thing like that, and he said, "They've only ruined one side of the mountain. Behind the pit, the range top is covered with snow. I can live with this. This is a part of the lifeblood of our nation." I recalled also that when the Beartooth Highway was built, ascending the wall of a Swiss-like valley to subsummit meadows of unique beauty, Love defended the project, saying that people who could not get around so well would be enabled to see those scenes.

Love is an unsalaried adjunct professor at the University of Wyoming, annually supervising the work of six or eight graduate students in Laramie and in the field. The imaginations of graduate students have a tendency to go dark when the time arrives to choose a topic for a thesis. Typically, they say to him, "Everything has been done."

"Nonsense," says the adjunct professor. "I can blindfold you and have you throw a dart at the geologic map of Wyoming, and wherever it hits you'll find a subject for a thesis."

One day in Jackson Hole, in the small log cabin Love sometimes uses as a field office, I asked him if I could throw a dart at the geologic map of Wyoming. "Be my guest," he said, and I sent one flying three times, scoring my first, third, and only Ph.D.s. The second one, for me, struck closest to home. It landed by Sweetwater Creek under Nipple Mesa, a couple of miles from Sunlight Peak in the North Absaroka Wilderness—eight miles from Yellowstone Park. "You have hit the Sunlight intrusives," Love said, and somehow I expected the sound of falling coins. "The area has not been surveyed," he continued. "There's no grid. Along Sweetwater Creek are mineral springs and oil seeps. A consortium of major oil companies wants the region removed from wilderness designation." The oil fields of the Bighorn Basin march across the sageland right to the feet of the Absarokas, he said, and their presence asks a great structural question: How far does the basin reach under the mountains? Since the Absarokas are made of volcanic debris, the oil seeping out of the banks of Sweetwater Creek could not have originated in Absarokan rock. He said he thought that the oil-bearing rock of the Bighorn Basin might go under the mountains all the way to Mammoth.

I repeated the name Mammoth, trying to remember where it was, and then said, "That's on the Montana border. It's all the way across Yellowstone Park."

He said, "Yes."

In 1970, Love and his colleague J. M. Good had published a paper on this subject. After considering and rejecting a number of titles—seeking to fashion the flattest and drabbest appropriate phrase—they settled upon "Hydrocarbons in Thermal Areas, Northwestern Wyoming." Now, with regard to my dart in the map, he said, "If you are interested in geochemistry, the composition of the oil from those seeps has not been studied. Is it Paleozoic high-sulphur oil? Mesozoic low-sulphur oil? Tertiary low-sulphur oil? One needs to know the quality of the oil and the depth of the reservoir rock." His tone seemed to exclude both emotion and opinion. "If you're interested in geophysics, what kinds of seismic reflections do you get from rocks below the volcanics?" he went on. "Can they be interpreted in a way that works out the prevolcanic structures? In terms of volcanic chemistry, what kinds of alteration of these Eocene volcanic rocks have occurred because of thermal activity and migration of oil into these rocks? None of this has ever been explored. In the regional context, a geologist cannot ignore the possibilities where that dart hit. A scientist, as a scientist, does not determine what should be the public policy in terms of exploration for oil and gas."

No rock could be more volcanic than the rock of Yellowstone Gorge—rose-and-burgundy, burnt-sienna, yellowcake-yellow Yellowstone Gorge—where petroleum comes out of the walls with hot water and steam. In 1939, when the National Park Service was digging abutments for a bridge downriver from the gorge, the National Park Service struck oil. Several workers, overcome by fumes of sulphur, died. These nagging facts notwithstanding, it was conventional wisdom in geology that where you found volcanics you would not find oil. In the nineteen-sixties, Love went out for a wider look. For example, he went on horseback into the Yellowstone backcountry carrying a four-foot steel rod. Twenty miles from the trailhead, he found swamps that were something like tar pits. When he jammed the rod into a swamp, a cream-colored fluid welled up. He put it in a bottle. In a day's time, the mixture had separated, and much of it was clear amber oil. In pursuing this project, the environmentalist within him balked, the user of resources preferred the resources somewhere else, but the scientist rode on with the rod. He knew he would bring scorn upon himself, but he was not about to stifle his science for anybody's beliefs or opinions. He did lose friends, including some Friends of the Earth. He lost friends in the Wilderness Society and the Sierra Club as well. To them, the Yellowstone oil was only the beginning of the threat he might be raising. The Designated Wilderness Areas of the United States had been selected on the assumption that they were barren of anything as vital as petroleum. "I will admit

that it bothers me that I have provoked the wrath of
organizations like the Sierra Club," Love remarked that
day in the cabin. "My great-uncle John Muir founded
the Sierra Club, and here I am, being a traitor."

Passing through Yellowstone on one of our journeys, Love and I found ourselves in foggy mists beside
a boiling spring, and on impulse he got out a scintillometer and held it over the water. The scintillometer
clicked away at a hundred and fifty counts per second,
indicating that the radioactivity in the spring was
about three times background. Interesting—but not
exactly adrenalizing to a man who had seen the thing
going at five thousand and upward.

In the years that immediately followed the Second
World War, the worldwide search for uranium was so
feverish that geologists themselves seemed to be about
three times background. Not only was the arms race
getting under way—with the security of the United
States thought to be enhanced by the fashioning of
ever larger and ever smaller uranium bombs—but also
there was promise of a panacean new deal in which this
heaviest of all elements found in nature would cheaply
heat homes and light cities. The rock that destroyed
Hiroshima had come out of the Colorado Plateau, and
it was to that region that prospectors were principally
drawn.

As any geologist would tell you, metal deposits were the result of hydrothermal activity. Geochemists imagined that water circulating deep in the crust picked up whatever it encountered—gold, silver, uranium, tin, all of which would go into solution with enough heat and pressure. They imagined the metal rising with the water and precipitating near the surface. By definition, a vein of ore was the filling of a fissure near a hot spring. This theory was so correct that it tended to seal off the conversation from intrusion by other ideas.

Three geologists working in South Dakota in 1950 and 1951 found uranium in a deposit of coal. Locally, there was no hydrothermal history. Oligocene tuff—volcanic detritus blown east a great distance—overlay the coal. There were people who thought that ordinary groundwater had leached the uranium out of the tuff and carried it into the coal. Love was one of the people. If such a process—contravening all accepted theory—had in fact occurred, then uranium might be found not only in hydrothermal settings but also in sedimentary basins. When Love proposed a search of Wyoming basins, hydrothermalists in the United States Geological Survey not only mocked the project but attempted to block it. So goes, sometimes, the spirit of science. The tuffs of the Oligocene were a part of the burial of the Rockies, and most had been removed during the exhumation. Love looked around for sedimentary basins where there was evidence that potential host rocks had once been covered with tuff. He had a DC-3 do surveys with an airborne scintillometer over the Powder River Basin. Some of the readings were remarkably hot—

notably in the vicinity of some high-standing erosional remnants called Pumpkin Buttes. He went there in a jeep, taking with him for confirming consultation the sedimentologist Franklyn B. Van Houten, who has described himself ever after as "Dave Love's human scintillometer." Love wanted to see if there had been enough fill by Oligocene time to allow the tuff to get over the buried Bighorn Mountains and be spread across the Powder River Basin. He and Van Houten climbed to the top of North Pumpkin Butte and found volcanic pebbles from west of the Bighorns in Oligocene tuff. Then Love went down among the sandstones of the formation lying below, where, at many sites, his Halross Gamma Scintillometer gave six thousand counts per second.

In time, he and others developed the concept of roll fronts to explain what he had found. In configuration, they were something like comets, or crescent moons with trailing horns—convex in the direction in which groundwater had flowed. As Love and his colleagues worked out the chemistry, they began with the fact that six-valent uranium is very soluble, and in oxidized water easily turns into uranyl ions. As the solution moves down the aquifer, a roll front will develop where the water finds an unusual concentration of organic matter. The organic matter goes after the oxygen. The uranium, dropping to a four-valent state, precipitates out as UO_2—the ore that is called uraninite.

One way to find deeply buried uraninite, therefore, would be to drill test holes in inclined aquifers. Wherever you found unusual concentrations of organic matter, you would move up the aquifer and drill again. If you found red oxidized sandstone, you would know that uraninite was somewhere between the two holes.

Drafting his report to the Geological Survey, Love described the "soft porous, pink or tan concretionary sandstone rolls in which the uranium was discovered," and added that "the commercial grade of some of the ore, the easy accessibility throughout the area, the soft character of the host rocks and associated strata, and the fact that strip-mining methods can be applied to all the deposits known at the present time, make the area attractive for exploitation." With those sentences he had become, in both a specific and a general sense, the discoverer of uranium in commercial quantity in Wyoming and the progenitor of the Wyoming uranium industry—facts that were not at once apparent. Within the Survey, the initial effect of Love's published report was to irritate many of his colleagues who were committed hydrothermalists and were prepared not to believe that uranium deposits could occur in any other way. They were joined in this opinion by the director of the Division of Raw Materials of the United States Atomic Energy Commission. A committee was convened in the Powder River Basin to confirm or deny the suspect discovery. All the members but one were hydrothermalists, and the committee report said, "It is true that high-grade mineral specimens of uranium ore were found, but there is nothing of any economic significance." Within weeks, mines began to open in that part of the Powder River Basin. Eventually, there were sixty-four, the largest of which was Exxon's Highland

Mine. They operated for thirty-two years. They had removed fifteen million tons of uranium ore when Three Mile Island shut them down.

In 1952, after Love's report was published, *The Laramie Republican and Boomerang* proclaimed in a banner headline, "LARAMIE MAN DISCOVERS URANIUM ORE IN STATE." The announcement set off what Love described as "the first and wildest" of Wyoming's uranium booms. "Hundreds came to Laramie," he continued. "I was offered a million dollars cash and the presidency of a company to leave the U.S.G.S. At that time, my salary was $8,640.19 per year."

The discovery predicted uranium in other sedimentary basins, and Love went on to find it. In the autumn of 1953, he and two amateurs, all working independently, found uranium in the Gas Hills—in the Wind River Basin, twelve miles from Love Ranch. By his description: "Gas Hills attracted everybody and his dog. It was Mecca for weekend prospectors. They swarmed like maggots on a carcass. There was claim-jumping. There were fistfights, shootouts. Mechanics and clothing salesmen were instant millionaires."

As it happened, he made those remarks one summery afternoon on the crest of the Gas Hills, where fifty open-pit uranium mines were round about us, and in the low middle ground of the view to the north were Muskrat Creek and Love Ranch. The pits were roughly circular, generally half a mile in diameter, and five hundred feet deep. Some four hundred feet of overburden had been stripped off to get down to the ore horizons. The place was an unearthly mess. War damage could not look worse, and in a sense that is what it was.

"If you had to do this with a pick and shovel, it would take you quite a while," Love said. The pits were scattered across a hundred square miles.

We picked up some sooty black uraninite. It crumbled easily in the hand. I asked him if it was dangerously radioactive.

"What is 'dangerously radioactive'?" he said. "We have no real standards. We don't know. All I can say is the cancer rate here is very high. There are four synergistic elements in the Gas Hills: uranium, molybdenum, selenium, and arsenic. They are more toxic together than individually. You can't just cover the tailings and forget about it. Those things are bad for the environment. They get into groundwater, surface water. The mines are below the water table, so they're pumping water from the uranium horizon to the surface. There has been a seven-hundred-per-cent increase of uranium in Muskrat Creek at our ranch."

We could see in a sweeping glance—from the ranch southwest to Green Mountain—the whole of the route he had taken as a boy to cut pine and cedar for corral poles and fence posts. An hour before, we had looked in at the ranch, where most of those posts were still in use—gnarled and twisted, but standing and not rotted. From John Love's early years there, when he slept in a cutbank of the creek, the ranch had belonged only to him and his family. The land was leased now—as was most of the surrounding range—to cattle companies. In the last half mile before we reached the creek, David counted fifty Hereford bulls and remarked that the lessees seemed to be overgrazing. "The sons of bitches," he said. "That's way too many for this time of year." Noticing some uranium claim stakes, he said, "People stake illegally right over land that has been deeded nearly a century."

Over the low and widespread house, John Love's multilaminate roof was scarcely sagging. No one had lived there in nearly forty years. The bookcases and the rolltop desk had been removed by thieves, who had destroyed doorframes to get them out. The kitchen doorframe was intact, and nailed there still was the board that showed John Love's marks recording his children's height. The green-figured wallpaper that had been hung by the cowboys was long since totally gone, and much of what it had covered, but between the studs and against the pine siding were fragments of the newspapers pasted there as insulation.

POSSE AFTER FIVE BANDITS

BATTLE NEAR ROCK ISLAND TRAIN

Robbers Are Found in Haystack
and Chase Becomes Hot

BOTH SIDES ARE HEAVILY ARMED

Fugitives Are Desperate, but Running Fight
Is Expected to End in Their Capture

Spinach had run wild in the yard. In the blacksmith shop, the forge and the anvil were gone. Ducks flew up from the creek. There were dead English currant bushes. A Chinese elm was dead. A Russian olive was still alive. David had planted a number of these trees. There was a balm of Gilead broadleaf cottonwood he had planted when he was eleven years old.

"It's going to make it for another year anyway," he said. "It's going to leaf out."

I said I wondered why the only trees anywhere were those that he and his father had planted.

"Not enough moisture," he replied. "Trees never have grown here."

"What does 'never' mean?" I asked him.

He said, "The last ten thousand years."

An antelope, barking at us, sounded like a bullfrog. Of the dozen or so ranch buildings, some were missing and some were breaking down. The corrals had collapsed. The bunkhouse was gone. The cottonwood-log granary was gone, but not Joe Lacey's Muskrat Saloon, which the Loves had used for storing hay. Its door was swinging in the wind. David found a plank and firmly propped the door shut. The freight wagon was there that he had used on trips for wood. It was missing its wheels, stolen as souvenirs of the Old West. We looked into a storage cellar that was covered with sod above hand-hewn eighteen-inch beams. He said that nothing ever froze in there and food stayed cold all summer. More recently, a mountain lion had lived there, but the cellar was vacant now.

In the house, while I became further absorbed by the insulation against the walls, Love walked silently from room to room.

Bizerta, Tunis, May 4—At a reception tendered him by the municipality, M. Pelletan, French Minister of Marine, in a brief speech, declared that France no longer dreamed of conquests, and that her resources would hereafter be employed to fortify her present possessions.

Cattle chips and coyote scat were everywhere on the floors. The clothes cupboards and toy cupboards in the bedroom he had shared with Allan were two feet deep in pack-rat debris.

Have you lost a friend or relative in the Klondike or Alaska? If so, write to us and we will find them, quietly and quickly. Private information on all subjects. All correspondence strictly confidential. Enclose $1.00. Address the Klondike Information Bureau, Box 727, Dawson, Y.T.

David came back into the space that had been his schoolroom, saying, "I can't stand this. Let's get out of here."

In the Gas Hills, as we traced with our eyes his journeys to Green Mountain, he said, "You can see it was quite a trek by wagon. Am I troubled? Yes. At places like this, we thought we were doing a great service to the nation. In hindsight, we do not know if we were performing a service or a disservice. Sometimes I think I might regret it. Yes. It's close to home."